Hemant Bessoondyal

Factors Affecting Mathematics Achievement

Hemant Bessoondyal

Factors Affecting Mathematics Achievement

The Mauritian Experience at Secondary Level

VDM Verlag Dr. Müller

Imprint

Bibliographic information by the German National Library: The German National Library lists this publication at the German National Bibliography; detailed bibliographic information is available on the Internet at
http://dnb.d-nb.de.

Any brand names and product names mentioned in this book are subject to trademark, brand or patent protection and are trademarks or registered trademarks of their respective holders. The use of brand names, product names, common names, trade names, product descriptions etc. even without a particular marking in this works is in no way to be construed to mean that such names may be regarded as unrestricted in respect of trademark and brand protection legislation and could thus be used by anyone.

Cover image: www.purestockx.com

Published 2008 Saarbrücken

Publisher:
VDM Verlag Dr. Müller Aktiengesellschaft & Co. KG , Dudweiler Landstr. 125 a,
66123 Saarbrücken, Germany,
Phone +49 681 9100-698, Fax +49 681 9100-988,
Email: info@vdm-verlag.de

Produced in Germany by:
Reha GmbH, Dudweilerstrasse 72, D-66111 Saarbrücken
Schaltungsdienst Lange o.H.G., Zehrensdorfer Str. 11, 12277 Berlin, Germany
Books on Demand GmbH, Gutenbergring 53, 22848 Norderstedt, Germany

Impressum

Bibliografische Information der Deutschen Nationalbibliothek: Die Deutsche Nationalbibliothek verzeichnet diese Publikation in der Deutschen Nationalbibliografie; detaillierte bibliografische Daten sind im Internet über http://dnb.d-nb.de abrufbar.

Alle in diesem Buch genannten Marken und Produktnamen unterliegen warenzeichen-, marken- oder patentrechtlichem Schutz bzw. sind Warenzeichen oder eingetragene Warenzeichen der jeweiligen Inhaber. Die Wiedergabe von Marken, Produktnamen, Gebrauchsnamen, Handelsnamen, Warenbezeichnungen u.s.w. in diesem Werk berechtigt auch ohne besondere Kennzeichnung nicht zu der Annahme, dass solche Namen im Sinne der Warenzeichen- und Markenschutzgesetzgebung als frei zu betrachten wären und daher von jedermann benutzt werden dürften.

Coverbild: www.purestockx.com

Erscheinungsjahr: 2008
Erscheinungsort: Saarbrücken

Verlag: VDM Verlag Dr. Müller Aktiengesellschaft & Co. KG , Dudweiler Landstr. 125 a,
D- 66123 Saarbrücken,
Telefon +49 681 9100-698, Telefax +49 681 9100-988,
Email: info@vdm-verlag.de

Herstellung in Deutschland:
Schaltungsdienst Lange o.H.G., Zehrensdorfer Str. 11, D-12277 Berlin
Books on Demand GmbH, Gutenbergring 53, D-22848 Norderstedt
Reha GmbH, Dudweilerstrasse 72, D-66111 Saarbrücken

ISBN: 978-3-639-06840-5

Science and Mathematics Education Centre

Gender and Other Factors Impacting on Mathematics Achievement at the Secondary Level in Mauritius

Hemant Bessoondyal

This thesis is presented for the Degree of
Doctor of Mathematics Education
of
Curtin University of Technology

October 2005DECLARATION

This thesis contains no material which has been accepted for any award of any other degree or diploma in any university.

To the best of my knowledge and belief this thesis contains no material previously published by any other person except where due acknowledgement has been made.

Signature: ....

Date: ....

ABSTRACT

Mathematics has been seen to act as a 'critical filter' in the social, economic and professional development of individuals. The Island of Mauritius relies to a great extent on its human resource power to meet the challenges of recent technological developments, and a substantial core of mathematics is needed to prepare students for their involvements in these challenges.

After an analysis of the School Certificate examination results for the past ten years in Mauritius, it was found that boys were out-performing girls in mathematics at that level. This study aimed to examine this gender difference in mathematics performance at the secondary level by exploring factors affecting mathematics teaching and learning, and by identifying and implementing strategies to enhance positive factors.

The study was conducted using a mixed quantitative and qualitative methodology in three phases. A survey approach was used in the Phase One of the study to analyse the performance of selected students from seventeen schools across Mauritius in a specially designed mathematics test. The attitudes of these students were also analysed through administration of the Modified Fennema-Sherman Mathematics Attitude Scale questionnaire.

In Phase Two a case study method was employed, involving selected students from four Mauritian secondary schools. After the administration of the two instruments used in Phase One to these selected students, qualitative techniques were introduced. These included classroom observations and interviews of students, teachers, parents and key informants. Data from these interviews assisted in analysing and interpreting the influence of these individuals on students, and the influence of the students' own attitudes towards mathematics on their learning of mathematics.

The results of Phases One and Two provided further evidence that boys were outperforming girls in mathematics at the secondary level in Mauritius. It was noted that students rated teachers highly in influencing their learning of mathematics. However, the teaching methods usually employed in the mathematics classrooms were found to be teacher-centered, and it was apparent that there existed a lack of opportunity for students to be involved in their own learning. It was also determined that parents and peers played a significant role in students' learning of mathematics.

After having analysed the difficulties students encountered in their mathematical studies, a package was designed with a view to enhance the teaching and learning of the subject at the secondary level. The package was designed to promote student-centred practices, where students would be actively involved in their own learning, and to foster appropriate use of collaborative learning. It was anticipated that the package would motivate students towards learning mathematics and would enhance their conceptual understanding of the subject. The efficacy of the package was examined in Phase Three of the study when students from a number of Mauritian secondary schools engaged with the package over a period of three months.

Pre- and post-tests were used to measure students' achievement gains. The What Is Happening in This Class (WIHIC) questionnaire also was used to analyse issues related to the affective domains of the students. An overall appreciation of the approaches used in the teaching and learning package was provided by students in the form of self-reports.

The outcomes of the Third Phase demonstrated an improvement in the achievement of students in the areas of mathematics which were tested. The students' perceptions of the classroom learning environment were also found to be positive. Through their self-reports, students demonstrated an appreciation for the package's strategies used in motivating them to learn mathematics and in helping them gain a better understanding of the mathematical concepts introduced.

ACKNOWLEDGEMENTS

I wish to express my gratitude to my supervisor, Professor John Malone, who has continuously provided his support, encouragement and help throughout this course. He has always been a source of inspiration for me.

My thanks also go to all the staff of the SMEC and friends who helped me when I was at the university for my residency. They were always ready to offer their help in times of need.

I wish also to thank the Mauritius Institute of Education for giving me the opportunity to pursue my studies and the staff of the Department of Mathematics Education for their constant support and encouragement. My thanks also go to the Ministry of Education and Scientific Research and the rectors of the different schools involved in this study for facilitating access to the schools and enabling the smooth conduct of the research.

This study would not have been possible without the contribution of students, teachers, parents and other stakeholders. I wish to thank all the people who were involved, directly or indirectly, in this study.

I have to make a special mention of my family members who provided the necessary emotional support and encouragement at all times.

TABLE OF CONTENTS

LIST OF TABLES

LIST OF FIGURES

shaded

LIST OF ABBREVIATIONS

ASEI: Activity, Student, Experiment and Improvise-

B.Ed: Bachelor of Education

B.Sc: Bachelor of Science

CPE: Certificate of Primary Education

DAST: Draw-A-Scientist-Test

ELP: Limited English Proficient

FIMS: First International Mathematics Study

HSC: Cambridge Higher School Certificate

IEA: International Association for the Evaluation of Educational Achievement

JICA: Japan International Cooperation Agency

MIE: Mauritius Institute of Education

NCTM: National Council of Teachers of Mathematics

PCEA: Roman Catholic Education Authority

PDSI: Plan, Do, See and Improve

PGCE: Post Graduate Certificate in Education

QTI: Questionnaire on Teacher Interaction

RDO: Research and Development Officer

SAC: Student centred, Activity based and Cooperative learning

SC: Cambridge School Certificate

SIMS: Second International Mathematics Study

SMASSE: Strengthening of Mathematics and Science in Secondary Education

SPSS: Statistical Package for Social Sciences

TIMSS: Third International Mathematics and Science Study

WIHIC: What Is Happening In this Class?

CHAPTER ONE
Context, Perspective and Overview

Mathematics in Education

Since the days of the early philosophers there has been a fascination with the beauty of mathematics and many of today's mathematicians believe that mathematics lies at the core of human knowledge. Mathematics continues as a pervading area of learning across the whole curriculum in schools around the world. The subject has a multiplicity of attributes ranging from the development of a person's logical reasoning to the understanding of abstract structures. It promotes logical and rational thinking and enhances one's ability to analyze and to solve problems. Life without mathematics is almost an impossibility and it would be difficult "to live a normal life in very many parts of the world" without it (Cockcroft, 1982, p.1). Its importance is also highlighted in the document entitled *Principles and Standards for School Mathematics* (NCTM, 2000, p.5) where the following appears: "those who understand and can do mathematics will have significantly enhanced opportunities and options for shaping their future". The main belief statement in this document is that all students should learn important mathematical concepts and processes *with understanding*. To achieve the vision for mathematics education described in that document requires "solid mathematics curricula, competent and knowledgeable teachers who can integrate instruction with assessment, education policies that enhance and support learning, classrooms with ready access to technology, and a commitment to both equity and excellence" (NCTM, 2000, p. 3).

In the 1970s Welsh (1978) observed several mathematics classes in operation.

> In all math classes that I visited, the sequence of activities was the same. First, answers were given for the previous day's assignment. The more difficult problems were worked on by the teacher or the students at the chalkboard. A brief explanation, sometimes none at all, was given of the new material, and the problems assigned for the next day. The remainder of the class was devoted to working on homework while the teacher moved around the room answering questions. The most noticeable thing about math classes was the repetition of this routine (cited in Romberg & Carpenter, 1986, p. 851).

Therefore, what can one assume about the teaching and learning of mathematics in classrooms today? Are students still passive receivers of a body of facts to remember and procedures to perform? Do they merely go through a considerable number of exercises in order to fill in the vacuum created by non-understanding of mathematical concepts and relationships? Do they just learn how to answer stereotyped, low cognitive level problems? Or, do students learn important mathematical concepts and processes with understanding? And, what is happening in mathematics classrooms in Mauritius? These sorts of questions were central to my study about mathematics classrooms in my country. This thesis describes the purpose, conduct, and findings of my study which examined the factors that affect the mathematics achievement of boys and girls at secondary level in Mauritius and how gender equity can be achieved.

In Mauritius, Lioong Pheow Leung Yung (1998) found that mathematics was being taught in a transmissionist, traditional way. Students were memorizing procedures and formulas; they were perfunctorily doing exercises, and lecturing was the most common method of teaching mathematics that they experienced. Importantly, ability differences between students were not addressed. As claimed by Griffiths (1998) mathematically capable students among those she studied in Mauritius were those who could adapt and respond to the answer-driven, drill-based system of mathematics education and they set the teaching pace in classrooms. As for the others, the less adaptive learners, their voices were rarely heard in the classroom — they were late in completing the class work and, often, the teacher was not even aware of their learning problems (Griffiths, 1998). Lioong Pheow Leung Yung's work (1998) pointed to the same problems at the secondary level in Mauritius where mathematics plays an important role in social and economic mobility. For example, in job applications, one of the prime criteria for employment that cuts across both the public and private sectors is the applicant's acceptable standard of mathematics and English. In other words, mathematics acts as a 'critical filter' whereby the pursuits of an individual can be constricted by a lack of mathematical knowledge. In view of the great importance given to mathematics in the school curriculum and in the economic and social mobility in society, a study in which the factors affecting the achievement of boys and girls is examined that has the potential to be very important.

Gender has been a matter of great concern in Mauritius and much is being done to ensure that gender equity and social justice exist. Several bills such as The Protection of the Child (Miscellaneous Provisions) Act (1998), The Sex Discrimination Act (2002) and The Protection from Domestic Violence (Amendment) Act (2004) have been passed in our Legislative Assembly in the recent years to deal with this issue. The concern of the government in the educational field can be felt through the following words of the Prime Minister of Mauritius in his message to secondary school students at the 34th Anniversary of the Independence, and at the 10th Anniversary of the Republic in March 2002.

> We are specially committed to promoting gender equity. We want more and more women to study science and technology so that they may equally contribute to the welfare of Mauritius.

The differences in the performance of Mauritian boys and girls in mathematics at the School Certificate level are discussed in a later section in this chapter. It suffices to say here that the achievement is inequitable; girls perform poorer than boys. In line with the government policy and the concern of all stakeholders in the educational field, a study that dealt with the issue of gender and mathematics was considered most important. Therefore, this study aimed to provide a research-based inquiry into the factors affecting the mathematics achievement of boys and girls at the secondary level in Mauritius. Besides achievement, other factors such as the role of students' attitude towards mathematics, teachers' interaction with and their expectations of students, and parental involvement and peer influence on the achievement of the students were examined. The focus of the examination was to analyze these factors from a gender perspective. Appropriate recommendations are made in this thesis for interested parties in the educational field in Mauritius. Hopefully the recommendations will enable them to address the problematic issue of gender's interaction with mathematics.

Precursors to the Study

It is important at the very outset in this thesis to discuss the background and relevance of this study to the issues discussed as well as my position on these issues.

My Personal Interests and Motivation

The eight years that I taught in a coeducational secondary school in Mauritius (1986-1994) provided me with the opportunity to note the strategies boys and girls adopt while learning mathematics and tackling mathematical problems. Over the past nine years, as a tutor in the mathematics education department of a teacher training institution, I have had further opportunities to examine more closely the issue of gender and mathematics. Over my professional life the different aspects of this issue have been discussed with teachers across Mauritius, and evidence of gender differences in mathematics achievement have been noted. I also had the opportunity to act as a supervisor for a number of dissertations for the Bachelor in Education

(B.Ed) and Post Graduate Certificate in Education (PGCE) courses that dealt with particular aspects of gender and mathematics. These experiences motivated me to plan for a study in which factors affecting mathematics achievement of boys and girls at secondary level in Mauritius could be examined. My ultimate aim was to identify ways and means to help boys and girls with their achievement in mathematics to ensure that the ideals of gender equity and social justice were met in their education.

International Concern about Gender and Mathematics

The problem of gender and mathematics has been an issue for many years and was recognized as a problem early in the nineteen century, as suggested by the following comments by Gauss in 1807:

> The taste for the abstract sciences in general and, above all, for the mysteries of numbers is very rare: this is not surprising, since the charms of this sublime science in all their beauty reveal themselves only to those who have the courage to fathom them. But when a woman, because of her sex, our customs and prejudices, encounters infinitely more obstacles than men in familiarizing herself with their knotty problems, yet overcome these fetters and penetrates that which is most hidden, she doubtless has the most notable courage, extraordinary talent and superior genius (cited in Leder, 1992, p. 597).

Since then much research have been carried out worldwide to deal with this issue. Overall, it has been found that girls participate less than boys do in mathematics and that there is 'a problem of girls and mathematics' (Willis, 1989). The 'problem' has been examined closely and a redefinition of it has been made as 'a problem of and with mathematics education, the way it is defined and the uses to which it is put' (Willis, 1989, p.2). The same remarks are made by Parker, Rennie, and Fraser (1996): "[T]he problem lies not with females themselves, but rather with the nature of science and mathematics and the presentation of these disciplines in school and society" (Parker, Rennie, & Fraser, 1996, p. xi).

It has been claimed that academic success in mathematics is mostly associated with males (Leder, 1989). Fennema (1974) has shown that there are differences between girls' and boys' learning of mathematics, particularly in activities that require complex reasoning. In another study, Fennema and Peterson (cited in Leder, 1989) concluded that students' achievement was influenced by their internal motivational beliefs, autonomous learning behaviours as well as external and societal factors. It also has been found that the pre-conceptions that girls hold concerning mathematics affect their attitude and performance in the subject. Numerous studies have concluded that teachers tend to interact more with boys than girls and direct higher ordered questions to boys than girls (Leder, 1996). Moreover, girls prefer more of a cooperative environment while boys prefer a more competitive environment (Fennema, 2000; Newquist, 1997). Furthermore, Rowe (1990) found that there are marked differences in the confidence levels in learning mathematics between boys and girls: boys are consistently more confident in their ability to do mathematics than girls. Fennema (2000) concluded that whenever higher-level cognitive skills are measured, girls are still not performing as well as boys, nor do they hold as positive an attitude toward mathematics. Fennema (1996) has stated:

> ...in spite of some indications that achievement differences are becoming smaller- and they were never large anyway – they still exist in those areas involving the most complex mathematical tasks, particularly as students progress to middle and secondary schools. There are also major differences in participation in mathematics-related careers. Many women, capable of learning the mathematics required, choose to limit their options by not studying mathematics (cited in Hanna, 1996, p.1).

A more detailed review of research conducted in this field will be carried out in Chapter Two.

Lack of Research Involving Gender and Mathematics in Mauritius

As already mentioned, gender issues are a matter of concern in Mauritius. Some studies have been carried out to examine the social aspects of the problem — for instance, Bunwaree (1996) highlighted the following concerning gender and education in the country:

> Mauritius has an official policy of equality of educational opportunity for boys and girls but this policy does not get translated into reality. Equality of opportunity does not only mean access to schools. Outcomes of schooling too are important in measuring equality (Bunwaree, 1996, p. viii).

She further argued that "schools do not act as agents of equality. Girls, especially from disadvantaged backgrounds, tend to be heavily discriminated against. Their academic performance tends to be poorer than that of boys" (Bunwaree, 1996, p.ix). Other studies in Mauritius also have been carried out on gender in education. For instance, the issues of girls' choice of physics and the involvement of female teachers at the primary level have been examined. It has been found that the enrolment and performance of students in Physics at the Higher School Certificate level both favour boys (Ramma, 2001). Concerning the teachers' involvement at the primary level in Mauritius, it was reported that there were more female primary teachers than males (Tengar, 2001). As far as gender and mathematics in Mauritius is concerned, some studies, as already mentioned, have been completed by students on this issue at the PGCE and B.Ed levels. However, a study to delve deeper into the gender issue problem in relation to mathematics education in secondary schools is imperative. It is important to examine the causes that affect the achievement of boys and girls and to identify ways and means to help different stakeholders in the educational field promote gender equity.

The Sociocultural Context of the Republic of Mauritius

As the study deals with factors affecting achievement in mathematics at the secondary level in Mauritius, it is appropriate that readers gain an idea of the background of the historical and social developments which have taken place in Mauritius. A brief account of the developments is provided below.

The Location of Mauritius and its Population

Mauritius is the main island of the Republic of Mauritius which consists of a number of small islands in the Indian Ocean scattered within a radius of 800 kilometres. The other islands are Rodrigues and Agalega. Mauritius covers an area of 1860 square kilometres and is approximately 800 kilometres from Madagascar and 2500 kilometres from Durban, South Africa. The island has no indigenous population and all the inhabitants are descendants of immigrants from different parts of the world. According to the Constitution of Mauritius, the population is composed of Hindu, Muslim, Chinese, and the General Population. There are a variety of languages in the islands: English, French, Creole, Mandarin, Hindi, Tamil, Telegu, Marathi, Urdu, and Bhojpuri.

Historical Developments in Mauritius

Historical records in Mauritius commence from the period when the Dutch first tried to settle in 1598, though it is claimed that Malay and Arab sailors, as well as the Portuguese, visited the island before that date (Ramdoyal, 1977). The Dutch named the country '*Mauritius*', after the Dutch Prince Count Maurice of Nassau (Bunwaree, 1994). Sugar cane was introduced in Mauritius by the Dutch and, after their departure in 1710, the French took possession of the island in 1715 and renamed it 'Isle de France'. Much development occurred on the Island and one can say that

Francois Mahe de Labourdonnais played a fundamental role in this achievement. As written by Bunwaree (1994) " ...he succeeded in transforming Ile de France from an insignificant outpost in the Indian Ocean into a prosperous French colony" (p.6). Other governors who succeeded Labourdonnais continued the developments until 1810 when the British conquered the island.

After the capture of the island on 3 December 1810, the British restored the Dutch name Mauritius. At the beginning, to ensure stability and continuity in the progress of the Island, the British opted to seek the cooperation of the French and economically powerful group, the sugar plantation owners. On 1st February 1835 slavery was formally abolished in Mauritius and, following the refusal of many of the freed slaves to continue working in the plantations, labourers mainly from India were brought to Mauritius. It should be mentioned also that other people from Fiji, South Africa, Sri Lanka, Trinidad and Guyana came to Mauritius. Under the British rule, sugar became the main export crop, and by 1850 the island 'emerged as the principal sugar producing colony' (Tinker, 1974; cited in Bunwaree, 1994, p. 13). Traders from China also settled in Mauritius and the Island continued to develop. On the 12th March 1968 it gained independence and then became a republic in 1992.

Educational Development in Mauritius

A review of the educational development that took place in Mauritius has been provided by Bunwaree (1994). During the time Mauritius was a French colony, education was an exclusive right of children of the French elite. A few attempts were made by some missionaries and others to provide education to the population, but without much success. An Ecole Centrale was created but it catered for the elite. The education provided had a heavy academic bias intended to assist those who wished to pursue university studies or take a professional career. The Ecole Centrale enjoyed a fine reputation through the qualities of the teachers and the system, and it even attracted students from other countries such as Portuguese Mozambique, the Cape of Good Hope and other parts of Asia (Bunwaree, 1994).

Since the children of coloured people were not given access to the Ecole Centrale, the Assembly and the Commission of Public Instruction approved the establishment of a separate school. It should be emphasized, however, that there were great disparities between the two schools:

> The stratified educational system placed the Ecole Centrale at the apex of the educational pyramid. While the external matters like class schedules,

examination times and dates were uniformly adopted, yet the content of the curriculum of the Ecole Centrale remained the monopoly of the elite that had access to it. Moreover, the competitive nature of the examinations further stratified even those candidates who attended the Ecole Centrale itself. (Prithipaul, 1976, p. 62)

Bunwaree (1994) described the system of education at the French period as follows:

The history of the French colonial period is characterized by the reluctance of the colonial administration to expand education on the island. There was strong discrimination against Blacks, Coloured and women (Bunwaree, 1994, p. 78).

When the British took over the island in 1810 there were some changes in the educational system, with the medium of instruction changing from French to English and the introduction of new subjects. However, admission to the Royal College was maintained for the children of the elite. Unequal opportunities in education were the concern of some people and Reverend Lebrun, a Roman Catholic priest, introduced free schools for children of coloured people. He contributed greatly to the cause of education of the coloured people and the slaves, and through his efforts and the support of the Secretary of State, the Royal College opened its admission to all citizens irrespective of race, colour, or class. It should be stressed that, in spite of the possibility of formal education from the government, the children of Indian immigrants received their education through informal institutions known as *baitkas, jammats,* and *madrassas.* These institutions played an important role in the preservation and dissemination of different cultural values and practices in Mauritius. Furthermore, the Church and missionaries have contributed much to the educational progress of the country. Grant-in-Aid, introduced in 1856, provided more opportunities for the Catholic Church in its endeavours to promote education. This system gave dual control to two educational provisions: one from the government sector and the other from the Church and the private sectors (Bunwaree, 1994).

The education system

The education system in Mauritius follows a British pattern with a three-stage model: 6+5+2 – that is, 6 years in primary (Standard I – Standard VI) leading to the Certificate of Primary Education (CPE); 5 years (Form I – Form V) in secondary school leading to the Cambridge School Certificate (SC); and 2 more years (Lower VI and Upper VI) in secondary school leading to the Higher School Certificate (HSC). Education is free in Mauritius from the primary to tertiary levels. Primary education has been made compulsory since 1993, secondary education made free since 1977, and tertiary education is free at the University of Mauritius (except for a nominal fee concerning registration and stationery). There are some courses at the masters level that are fee paying. Also, some fee-paying institutions exist at the primary, secondary and tertiary levels.

The diagram below shows the formal educational system in Mauritius.

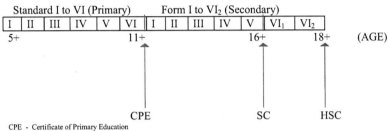

CPE - Certificate of Primary Education
SC - Cambridge School Certificate
HSC - Cambridge Higher School Certificate
VI₁ - Six One (Lower Six or first year HSC)
VI₂ - Six Two (Upper six or second year HSC)

According to the Digest of Educational Statistics (2003) there were 291 primary schools in Mauritius in 2003 of which 221 are administered by the government and 51 by the Roman Catholic Education Authority (RCEA), two by the Hindu Education Authority and the rest were private unaided schools. A total of 129 616 pupils (66 104 boys and 63 512 girls) were enrolled in the primary school of which 76% were enrolled in the government schools. There were 4 247 General Purpose Teachers (1 793 males and 1 934 females) and 1 373 Oriental Language Teachers (748 males and 740 females). Students aged 5+ enter the primary school usually after at least one year of pre-primary schooling. They spend six years of compulsory education, from Standard I to Standard VI, and then take the Certificate of Primary Education (CPE). A total of 27 842 students took the examinations in 2002 and the pass rate was 65% (60% boys and 71% girls). Successful students then move to the secondary schools, based on their performance in the CPE examinations.

The corresponding statistics for the secondary sector for 2003 reveal 175 secondary schools, of which 63 are State Secondary Schools and the rest are Private Secondary

Schools (including Confessional Schools; that is schools run by religious bodies). In 2003 there were 103 847 students (49 946 boys and 53 901 girls) enrolled in secondary schools, of which 30% were in the state schools. There were 2 104 teachers (975 males and 1 129 females) employed in the State Secondary Schools whereas the private schools employed 3 834 teachers (1 896 males and 1 938 females). Teachers in the secondary sector have qualifications ranging from School Certificates and Teacher's Certificates for those teaching lower forms, to degree holders and those with professional qualifications teaching up the HSC level. Teachers at the HSC level teaching mathematics are predominantly males, but in recent years an impressive number of female teachers holding degrees are teaching the subject at this level.

Table1.1 illustrates the highest qualifications of the teachers in secondary schools as at 2003.

Table 1.1: Highest qualifications of teachers in secondary schools in 2003

	School Certificate or equivalent	Higher School Certificate or equivalent	Certificate or Diploma	First Degree	Post Graduate	Not stated	Total
Males	146	397	577	1510	238	3	2871
Females	107	517	650	1577	207	9	3067
Total	253	914	1227	3087	445	12	5938

Digest of Educational Statistics (2003)

Successful students from the primary school level spend their first five years (from Form I to Form V) preparing for the Cambridge School Certificate (SC) examination (O-Levels). Of the 14 527 students who took the examinations in 2002, 75% passed (72% boys and 77% girls). Students who successfully complete the SC examinations have the possibility of studying for two further years, at the end of which they sit for the Cambridge Higher School Certificate (HSC) examination (A-Level). Of the 6 845 students who took the HSC examination in 2002, 76% passed (74% boys and 77% girls).

The best candidates at the HSC examinations (Laureates) are awarded State Scholarships for higher studies abroad. This makes the whole system a highly competitive one and consequently there is a considerable emphasis on drill work for examinations. This has unfortunately led to the running of a parallel system of education in the form of private tuition which, is in fact, almost institutionalised from the commencement of the primary level. It is really hard to find a secondary student in the higher forms who is not studying under private tuition.

A pre-vocational stream has been set up to accommodate the primary-school leavers who fail to obtain a CPE. As at March 2003 there were 114 schools offering pre-vocational education, of which 49 were state schools while the rest were confessional or private schools. It should be noted that only 9 schools offer pre-vocational education, while the remaining 105 offer both secondary academic and pre-vocational education. In 2003 there were 7 326 students (4 673 boys and 2 653 girls) enrolled for pre-vocational education of which 37% were in the state schools. The pre-vocational education teaching staff was 433 (191 men and 242 women).

Development of Education for Girls in Mauritius

According to historical records, the first female school was set up during Decaen's period (1802-1810) as Governor of the colony (Bunwaree, 1994). The curriculum of the school was geared towards shaping the girls for societal acceptance. Music, drawing, French grammar, and sewing were some of the subjects taught. The contribution of Deaubonne in female education in Mauritius has been highlighted:

> …who with his family has founded a school for young girls and has thus filled a vacuum in the life of the history. The school serves a useful purpose in a country where the youths had no other example but the depraved lives of their slaves, who were the only ones rashly charges by the parents to take care of them. (Tombe, 1810, cited in Ramdoyal, 1977, p. 27-28.)

The one female school run by Deaubonne closed in 1809. It can be said that girls' education in Mauritius was encouraged with the opening of Loreto schools by religious institutions (Caunhye, 1993). It should also be pointed out that many other cultural associations were contributing in their own way to the education of the girls in their community.

In 1949, the first Mauritian girl was awarded an English scholarship based on her performance in the Higher School Certificate Examination, and during 1959 it was decided to increase the number of English scholarships for girls to two. The number of these scholarships awarded to boys was four. At primary level, in 1950, 15 scholarships were awarded to girls and 30 to boys. At the same time, there were four scholarships open to girls and six to boys other than those in the government or Aided primary schools who were qualified in respect of age. The number of Junior scholarships was raised to 120 in 1957: 70 for boys and 50 for girls (Caunhye, 1993).

Free education in 1977 contributed considerably to the education of girls in Mauritius. Before that, secondary education was fee paying and many girls were 'sacrificed' to allow their brothers to continue secondary education. Some could not attend secondary schools because of financial constraints, some were needed to look after younger brother or sister and some because of household obligations. When secondary education became free, the problem of financial constraint was solved and this resulted in an increase of girls' involvement at secondary level and eventually even at the tertiary level. The trend in enrolments of males and females at the secondary level in Mauritius is shown in Figure 1.1.

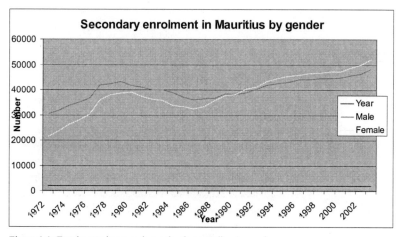

Figure 1.1: Enrolments in secondary schools according to gender

From the graph drawn in Figure 1.1 it can be noted that the enrolments of both boys and girls have increased over time with those of girls having more than doubled in the thirty years (1973-2003). It can also be noted that girls' involvement in secondary education surpassed that of boys in the early 1990's.

The Theoretical Framework of the Study

The main conceptual referent in the study is the issue of equity in education. This issue has been of major concern in our country. Primary schooling is compulsory in

Mauritius and recently The Education (amendment) Act (2004) has been passed in the National Assembly to make secondary education compulsory (academic or vocational) until students are 16 years old. The Sex Discrimination Act (2002) has been passed in the National Legislative Assembly and in the last Budget Speech (2004-2005) the Equal Opportunities Bill was announced to help assure equity in the society. It has been established that every child who has finished the Certificate of Primary Education should be admitted to a secondary school. The question is whether this action would bring equity in education: would the different facilities, learning environments and attitudes of students to education in the different types of schools hamper the establishment of equity in education? Equality of opportunity is an important ingredient for gender and education and Robertson (1998, p. 188) rightly pointed out:

> Giving all children the opportunity to enjoy an equal education, determined
> not by the wealth of their families, but by the resources of their communities,
> is … a truly democratic ideal… a shared public commitment in achieving
> greater equity is the only reason for public schools to school.

Equality of opportunity is definitely a necessary component for equity in education, but it is not sufficient. Attempts to achieve equality of opportunity for boys and girls have focused mainly on equality of resources and access to curriculum offerings. Yates (1985) made the point clearly.

> A simple policy of equal offerings was a naïve approach to social equality
> given the different out-of-school experiences of girls and boys…where the
> criteria of success and the norms of teaching and curriculum are still defined
> in terms of the already dominant group, that group is always likely to remain
> one step ahead. (cited in Gipps & Murphy, 1994, p.10)

It was reported (Gipps & Murphy, 1994, p.11) that in the United Kingdom (UK), even almost 20 years after the passing of the Sex Discrimination Act 1975, there was still a problem of assuring equal educational chances for girls and boys. Different expectations and interactions of teachers, among other factors, do contribute in this problem. While discussing equity, Secada (1989) mentioned that "[e]ducational equity should be construed as a check on the justice of specific actions that are carried out within the educational arena and the arrangements that result from those actions" (cited in Gipps and Murphy, 1994, p.13). The focus has now shifted from equality of opportunity to equality of outcomes. Fennema (2000, para. 6) argues that

Instead of equal educational experiences, equity could mean equality of outcomes, i.e. that females should learn exactly the same mathematics as do males, be able to perform the same on various measures of mathematical learning, and have the same personal feelings towards oneself and mathematics.

It is believed that when this equity is achieved, girls will be as confident about learning mathematics as boys; girls' overall attitude towards mathematics will have changed, and there would be no differences in mathematics achievement (both in quantity and quality) between boys and girls. This study has attempted to analyse the extent to which Mauritius has achieved this aspect of equity, and what should be done further to ensure continued gender equity and social justice.

Purpose of the Study

The purpose of this study was to identify the problems that boys and girls encounter in their learning of mathematics at the secondary level in Mauritius and also factors that impact on their mathematics achievement. Several sociocultural influences were examined in this study: the attitude of boys and girls towards mathematics; their motivation towards the subject; their participation in classroom discussion; the type of questions they ask and are asked by the teacher; the type of teacher interaction with boys and girls in mathematics classes and the involvement of parents in their children's education. I believe that there may be other factors affecting mathematics achievement of students at the secondary level and a number of these became evident as the study progressed – as discussed in Chapter Five. A teaching and learning package to help in enhancing the mathematics achievement and attitude towards mathematics of boys and girls at the secondary level in Mauritius was also designed and tested in this study. Action to address gender equity was identified and proposed to the different stakeholders in the educational society of Mauritius.

In order to provide a focus for these aims the following research questions were formulated:

1. What are the factors that contribute to the mathematical achievement of Mauritian students at the secondary school level?
2. What types of difficulties do Mauritian boys and girls encounter while learning mathematics at the secondary level?
3. Why do these difficulties occur?

4. How effective is a teaching and learning package based on the findings of (1) above at enhancing the attitudes and mathematical achievement of secondary school students?

The Research Design

The study was conducted using a mixed mode, where both quantitative and qualitative methods were used. It involved questionnaires, case studies, classroom observations and interviews. The methodology used in this study is described in detail in Chapter Three. The research design for the study consists of three phases. Before actually starting with the first phase, an examination of the current educational system in Mauritius was made in order to establish a criteria on which the participating schools were to be chosen. The zoning system present in Mauritius and the type of school played a dominant role in this choice (more details will be given in Chapter Three). Other documents also were analyzed in order to design a survey questionnaire dealing with mathematical problems based on the school curriculum. After the questionnaire was developed it was piloted, and based upon the feedback obtained, the questionnaire was amended.

The first phase of the study consisted of administering the problem solving questionnaire, along with a second questionnaire focusing on the attitude of students to mathematics, to a sample of students of Form Four (average age 15 years old) chosen from the seventeen secondary schools. The completed questionnaires were analyzed to obtain an overall picture of the issue of gender and achievement in mathematics in the Republic of Mauritius. The outcomes of this phase helped in structuring the second phase of the study.

In the second phase, a case study approach was used. A representative sample of four schools was chosen: one single-boys school, one single-girls school, and two coeducational schools. A class of Form Four (average age 15 years old) in each of the schools was then chosen and the questionnaires were administered. After analyzing data collected through the questionnaires, appropriate issues were identified to be probed further during interviews with the students. Importantly, during this phase classroom observations were carried out in the four schools involved. The participation of students, type of intervention, frequency of intervention, reaction of the teacher to the students, and other related issues were observed. The Questionnaire on Teacher Interaction (QTI) was also administered and results were analysed to probe into the type of interactions between the teachers and the students in the classrooms. With a view to obtain the perception of the students towards mathematics and mathematicians, the Draw-A-Mathematician-Test (Picker & Berry, 2000) was administered to all the students involved in this sample. The students were also asked to explain their drawings and provide further clarifications about their images. This phase of the study also included interview with teachers, rectors, parents, and other stakeholders in the educational field. Based on the data collected from the different techniques and statistical analysis conducted, appropriate measures to respond to the issue of gender and mathematics at secondary level in Mauritius were identified.

The third phase of the study involved the implementation and evaluation of a teaching and learning package based on recommendations identified in the second phase. For this purpose, three secondary schools (one single-boys school, one single-girls school and one co-educational school) were chosen. One Form IV class from each of these schools was identified and a pre-test relating to the mathematics concepts they were assumed to know was conducted. Appropriate strategies were implemented by me in each of the classes with a view to enhance the teaching and learning of mathematics at that level for a period of three months. After that, post-tests were administered to the students to evaluate the effectiveness of the strategies used. Learning environment questionnaires followed by interviews were also administered to obtain feedback on whether there had been a change in attitude towards mathematics among students.

Significance of the Study

This study is significant for two major reasons. First, the evidence in Mauritius is clear that the achievement of boys and girls in mathematics at secondary schools is inequitable. The study should throw some light on why this is so. Second, the study will provide information for educational stakeholders to plan appropriate action to address the imbalance between boys and girls mathematics achievement. Consider the following evidence in addressing the first issue.

Achievement of Boys and Girls in Mauritius

The results of boys and girls at the first national examinations (end of primary schooling, average 10-11 years old) are provided in Tables 1.2 and 1.3.

Table 1.2: Percentage of Passes in Certificate of Primary Education Examination (CPE) in Mauritius 1995-2004

	1995	1996	1997	1998	1999	2000	2001	2002	2003	2004
Boys	60.4	60.0	59.4	59.9	57.5	58.5	56.6	57.7	53.8	54.7
Girls	67.7	67.7	67.3	67.3	67.4	69.9	68.9	69.3	66.9	65.8

Source: Mauritius Examination Syndicate (2005)

It can be noted that girls consistently performed better than boys in the end-of-primary national examinations. To gain an insight of the situation in mathematics at the level, the percentage of passes of boys and girls in mathematics are presented in Table 1.3.

Table 1.3: Percentage of Passes in Mathematics in CPE Examination in Mauritius 1995-2004

	1995	1996	1997	1998	1999	2000	2001	2002	2003	2004
Boys	69.6	70.1	70.5	71.4	69.5	69.9	68.8	71.2	69.9	70.3
Girls	74.3	74.5	75.1	76.9	75.7	77.4	77.2	78.3	78.8	77.8

Source: Mauritius Examination Syndicate (2005)

Girls are still performing better than boys in the subject of mathematics at the primary school level. What is the situation at the School Certificate level? Table 1.4 shows the result of boys and girls in the School Certificate Examination for the past ten years.

Table 1.4: Percentage of Passes in School Certificate Examination in Mauritius 1995-2004

	1995	1996	1997	1998	1999	2000	2001	2002	2003	2004
Boys	69.2	72.8	74.2	77.6	76.3	75.3	75.1	72.8	73.5	75.9
Girls	72.0	77.6	77.8	78.1	78.6	79.4	79.8	77.3	78.4	80.1

Source: Mauritius Examination Syndicate (2005)

Based on the data in Table 1.4, one can observe that for the School Certificate Examinations, the tendency is for girls to perform better than boys at the secondary level. What is the situation concerning mathematics? The percentage of passes in mathematics for the past ten years are displayed in Table 1.5.

Table 1.5: Percentage of Passes in Mathematics in School Certificate Examination in Mauritius 1995-2004

	1995	1996	1997	1998	1999	2000	2001	2002	2003	2004

| Boys | 75.8 | 76.7 | 75.1 | 73.0 | 73.5 | 79.3 | 79.1 | 70.3 | 71.9 | 74.8 |
| Girls | 69.3 | 72.0 | 71.6 | 68.9 | 66.2 | 70.5 | 71.5 | 66.9 | 68.6 | 70.6 |

Source: Mauritius Examination Syndicate (2005)

The difference and the trend in these data can be easily read from the line graphs in Figure. 1.2.

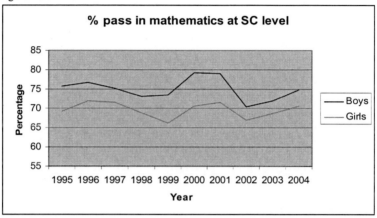

Figure 1.2: Percentage of Passes in Mathematics at SC Level

It is very clear that in Mauritius the performance of secondary girls in mathematics is poorer than that of boys. The same trend can be observed when the grade distribution is analyzed.

Table 1.6: Percentage of Students Scoring Grades 1-3 in Mathematics in the School Certificate Examination 1995-2004

	1995	1996	1997	1998	1999	2000	2001	2002	2003	2004
Boys	29.6	31.0	30.3	27.2	27.9	27.8	25.7	24.6	23.8	27
Girls	20.9	21.7	22.7	19.2	20.8	20.7	20.4	21.3	21.4	23.8

Source: Mauritius Examination Syndicate (2005)

A graphical presentation of the data is provided in Figure. 1.3.

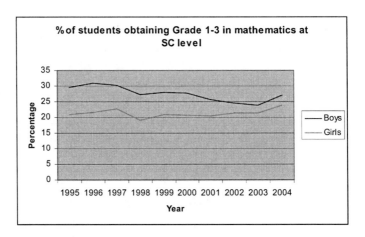

Figure 1.3: Percentage of Students Scoring Grades 1-3 in Mathematics in the School Certificate Examination

In the (1995-2004) achievement data there is a significant difference in the performance of boys and girls in mathematics at the secondary level (though there is recent evidence that the gap is diminishing). An extensive review of the literature has revealed that little research has been carried out in relation to gender differences in mathematics in Mauritius except for a few dissertations at the PGCE or B.Ed level. It was necessary to probe deeper into this research problem. The factors responsible for such a difference needed to be identified and appropriate responses proposed.

Action for Educational Stakeholders in Mauritius

Perhaps more importantly, this study is significant because it will provide different stakeholders in Mauritius involved in the education sector (policy makers, curriculum developers, teacher trainers, secondary school teachers, rectors and parents) with helpful findings concerning gender and mathematics. Appropriate responses and action could then be taken by these stakeholders in accordance with the country's policy of gender equity and distributive justice.

Overview of the Thesis

The purpose of this chapter has been about positioning the sociocultural context of the study and providing a broad outlook of the problem and the research questions that guided the study. The theoretical framework and significance of the study also have been discussed. Chapter Two examines relevant literature concerning the issue of gender and mathematics and other related concepts. The conceptual framework chosen for the study is also discussed in this chapter. Chapter Three describes the grounded theory methodology underpinning the study and the three stages in the research design adopted for the study. The associated issues concerned with the methodological framework, triangulation, standards in research, and ethical issues are described in this chapter along with the different criteria used for selecting the sample at each stage, the instruments used to collect data, and the survey questionnaires designed for the specific purpose of this study.

The analysis of the data obtained from the two questionnaires administered in the first phase of the study constitutes Chapter Four. Statistical techniques were used to analyze possible correlations between the different factors identified in mathematics achievement with a gender perspective. Qualitative techniques also were used to probe deeper into the issues. Chapter Five documents the different observations carried out within the classrooms of the four secondary schools chosen for this stage. The analysis of the data obtained from the two questionnaires administered to these three schools, the QTI and the Draw-A-Mathematician-Test are also discussed. Interviews with students, teachers, parents, rectors, and other stakeholders form an additional part of the Chapter Five. The first three research questions will then be answered based on the findings of phases one and two of the study.

Chapter Six deals mainly with the third phase of the study, namely the implementation phase. The strategies and recommendations identified are described in Chapter Six together with how the lessons were conducted in three secondary schools to test their efficiency. The questionnaires that were used, the framework in which the lessons were devised together with the response of the students also form part of this chapter. After analyzing the performance of the students in the pretest and posttest, and analyzing the responses of the students in the WIHIC questionnaires and in the interviews, the fourth research question will be answered. In Chapter Seven the main findings from the different phases will be synthesized using a thematic approach. The factors affecting mathematics achievement of boys and girls, the way these factors interact, the difficulties our children in Mauritius encounter while learning mathematics at secondary level and why these difficulties occur are all discussed in this chapter. Appropriate recommendations are proposed to respond to the issue of gender and mathematics and ensure gender equity in our classrooms and educational system. The limitations of the study are also discussed and they conclude the thesis.

The following chapter reviews literature related to this study and highlights where the present study contributes to, and adds to, the literature base.

CHAPTER TWO
Literature Review

As explained in Chapter One, the main aims of this study were to determine a number of factors that impact on the mathematics achievement of boys and girls in Mauritius; discuss the difficulties the students encounter while learning mathematics; determine the reasons for these problems, and propose pedagogical strategies through a teaching and learning package to help students in their learning of mathematics.

To identify the factors that need consideration, previous related studies that were conducted in different parts of the world were reviewed. These were examined with a view to provide a focus for this study and for the research questions. This chapter describes the literature review and situates the relevance of the previous research to this study. The contributions of the present study to the body of literature are also discussed in this chapter.

It was noted in Chapter One that girls perform better than boys in mathematics at the end of the primary examination in Mauritius, while at the School Certificate level it is the boys who are performing better. In order to devise strategies to assist all students to achieve well in such an important subject as mathematics (for further education or job opportunities), a review of studies conducted in relation to gender and mathematics were regarded as important.

After having taken cognizance of the different issues concerning gender and mathematics, it was also considered that the focus of the study should be on Mauritius. The proposed strategies could be worked upon and their effectiveness in the Mauritian society evaluated through an implementation program. Consequently, this chapter commences with major findings in the field of gender and mathematics. Learning environment plays a very important role in enhancing the teaching and learning process and thus this chapter also includes one section where the different aspects of learning environment and a number of learning environment and other questionnaires are discussed. The questionnaires administered in this study are also identified in this section. Research (Khalid, 2004; Schiefele & Csikszentmihalyi,

1995) have shown that one way to bring about changes in the teaching and learning of mathematics is to arouse the interest of the students in the subject and motivate them towards their learning. Consequently, studies relating to motivation and mathematics were reviewed in order to help me in planning the strategies to be used in the implementation stage of this study. These are incorporated in a section of this chapter where studies related with teaching of mathematics with understanding are also dealt with.

In Chapter One it was noted that students in Mauritius do encounter problems in mathematics at the secondary level and the way the subject is normally taught was discussed upon: it was noted that the teaching was transmissionist and was carried out in a traditional way. Lecturing was the most common method of teaching mathematics and emphasis was more on an algorithmic approach to solve problems. With a view to bring changes in the current way of teaching the subject, a review of the studies related to teaching mathematics for conceptual understanding was considered important and this was included in the chapter along with a description of the conceptual framework of the study and a concluding summary.

Gender and mathematics

Gender and mathematics has been the focus of considerable research over the past thirty years since 1974 (Becker, 2003; Burton, 1990; Ercikan, McCreith, & Lapointe, 2005; Ethington, 1990; Fennema, 1974, 1990, 1995, 1996, 2000; Fennema, Carpenter, Jacobs, Franke, & Levi, 1998; Fennema & Leder, 1990; Fennema & Sherman, 1977; Fennema & Tartre, 1985; Forbes, 1999; Hanna, 1986, 2003; Hanna, Kundiger, & Larouche, 1990; Hanna & Nyhof-Young, 1995; Leder, 1982, 1987, 1989, 1990a, 1990b, 1990c, 1992, 1995, 1996; Leder & Fennema, 1990; Leedy, LaLonde, & Runk, 2003; Sprigler & Alsup, 2003; Zevenbergen & Ortiz-Franco, 2002). Gender related differences in mathematics achievement have been reported in many studies where boys were performing better in mathematics than girls (Hanna et al., 1990; Köller, Baumert, & Schnabel, 2001; Leder, 1992; Seegers & Boekaerts, 1996).

The data from the Second International Mathematics Study (SIMS) was analysed by Hanna (1986) to examine gender differences in the mathematics achievement of

Canadian eighth grade students. Five areas were surveyed: arithmetic, algebra, probability and statistics, geometry, and measurement. Concerning the first three areas, no differences were found in the performance of boys and girls. However, for geometry and measurement the boys' mean percent of correct responses was higher than for girls and these differences were found to be statistically significant at the 0.01 level. Hanna, Kundiger & Larouche (1990) also analysed the 1982 SIMS data. In a study that focused on Grade 12 students from fifteen countries. They found that whenever gender differences were identified, they were in favour of boys and that it was only in Thailand where no significant difference was noted.

A summary of the gender analysis of three studies conducted within the space of 30 years by the International Association for the Evaluation of Educational Achievement (IEA), namely the FIMS, the SIMS and the TIMSS is provided in Hanna (2003, p. 210) and shown in Table 2.1.

Table 2.1: Gender Analysis of FIMS, SIMS and TIMSS

		Age 13		Ages 17-18
FIMS **(1964)**	1.	*Differences in boys' favor in 10 out of 12 countries.*	1.	*Differences in boys' favor in all 10 countries.*
	2.	*Considerable variation between countries in the extent of gender differences.*	2.	Considerable variation between countries in the extent of gender differences.
SIMS **(1980-82)**	1.	*No difference in 5 out of 20 countries on all subjects.*	1.	*No differences in 3 out of 15 countries on 6 out of 7 subtests.*
	2.	*Differences in boys favor in 10 countries, in up to 2 out of 5 subtests.*	2.	Differences in boys' favor in 12 countries on 2 to 6 subtests.
	3.	*Differences in girls' favor in 5 countries in up to 2 out of 5 subtests.*		
TIMSS **(1995)**	1.	*No differences in overall achievement in 37 out of 39 countries.*	1.	*No differences in 5 out of 16 countries.*
	2.	Slight differences in girls' favor in Algebra in 12 countries (in Grade 8).	2.	Differences in boys' favor in 4 countries on up to 2 content areas and in 7 countries on each of the 3 content areas.

As a result of these studies, different forums were constituted, conferences organized and consciousness raised concerning equity in mathematics education around the world. Different intervention programmes were developed in view of increasing females' enrolment and attainment in mathematics and mathematics oriented courses (Fennema, 2000; Hanna, 2003; Leder, 1992). A number of studies conducted afterwards showed that in many countries the gender differences in enrolment, and even performance, have decreased and are now quite small (Elwood, 1999; Fennema, 1996; Friedmann, 1989; Hanna, 2003). However other studies did find that even if the gender differences gap in mathematics achievement have been shrinking over time, gender differences favouring males still existed in areas related to high-level cognitive skills (Casey, Nuttall, & Pezaris, 2001; Fennema, 2000; Hyde, Fennema, & Lamon, 1990; Leder, 1992). In another study, Johnson (1996; cited in Köller et al., p. 452), after having summarized several large-scale national and international assessments of mathematics achievement, concluded that "the evidence is clear and overwhelming...that achievement differences in favour of boys exist in mathematics...". More recently Köller et al. (2001, p. 452) have confirmed that girls in Germany still do not perform as well as boys in mathematics. There are also counter-claims that now the boys are at the disadvantageous position (Boaler, 1997; Hanna, 2003; Matters, Allen, Gray, & Pitman, 1999; Vale, Forgasz, & Horne, 2004).

As mentioned earlier, in Mauritius there is still a difference in the mathematical performance of boys and girls of average age 16 years, with boys performing better. However, there has been a slight decrease in this difference, starting when the students joined the secondary schools as shown by the data presented in Table 1.5 on Pg. 17. The results of mathematics at the end of primary level show that at that stage the performance of girls is better than boys. The present study aims at analyzing why this is so and producing some suggestions for reaching gender equity.

Three major trends were identified by Dunne and Johnson (1994) relating to studies concerning gender and mathematics/science education. One strand was mainly devoted to finding and documenting differences. Alternative methods were used to compare enrolments and achievements of boys and girls in mathematics. One of the first concerns related to gender and mathematics was realised in the 1970's, namely the marked difference in enrolments between males and females (once it ceased to be compulsory). Leder (1990b) noted that the enrolments for more advanced mathematics courses such as trigonometry, pre-calculus and calculus, indicated fewer females than males in the US were enrolling for these courses. She mentioned that a similar trend could be observed in other countries also. Sanders and Peterson (1999) asserted that the experiences girls have had in middle school and high school mathematics classes often affects their enrolments and caused negative attitudes towards mathematics at the post-secondary level.

Mathematics plays an important role in the enrolment in many fields of study or job opportunities in Mauritius. A poor performance in mathematics results in opting out of many courses and consequently, opting out of many job opportunities. As stated by Schoenfeld (2002): To fail children in mathematics, or to let mathematics fail them, is to close off an important means of access to society's resources". In other words, it acts as a critical filter in the educational and social mobility of people. Marked differences in enrolments in mathematics-related courses were resulting in gender differences in certain positions of the Mauritian job market. The following view about mathematics in society is worth noting:

> Mathematics is widely recognised not only as a core component of the curriculum but also as a critical filter to many educational and career opportunities. Yet in recent years much concern has been expressed about students' reluctance to continue with the study of mathematics well beyond the compulsory years, a trend often described emotionally as the drift away from mathematics and from the physical sciences.
>
> (Leder, Pehkonen, & Torner, 2002, p. 1)

To provide an indication of the gender differences in enrolments among Mauritian students in mathematics related subjects, the intake rate at the University of Mauritius in Mathematics and Engineering courses over a five year period is given in Table 2.2.

Table 2.2: Enrolment in BSc mathematics at the University of Mauritius

Year	1999-2000	2000-2001	2001-2002	2002-2003	2003-2004
Male	17	14	9	11	10
Female	25	27	16	15	15

These enrolments are shown using a line graph in Figure. 2.1.

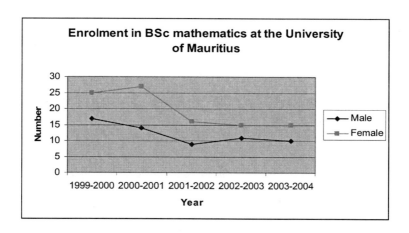

Figure 2.1: Enrolment genderwise in BSc mathematics at the University of Mauritius

It is to be noted that the girls consistently outnumber boys in the enrolment for the BSc in mathematics at the University of Mauritius, but one should note that both these numbers are very small. In fact, data suggest that in the 1990's, out of 100 students passing the School Certificate examination, 44 students joined Lower VI, 32 passed the Higher School Certificate examination and 18 joined the tertiary level. The present situation is that out of 100 students passing the School Certificate, 65 joined Lower VI, 48 passed Higher School Certificate examination and 27 joined the tertiary level. It should also be noted that more students now are continuing education but the structure of the education system is still pyramidal.

To have an idea of the involvement of boys and girls in a mathematics related course, the enrolments of boys and girls in engineering for the past five years at the University of Mauritius are given in Table 2.3.

Table 2.3: Enrolment in Engineering at the University of Mauritius

Year	1999-2000	2000-2001	2001-2002	2002-2003	2003-2004
Male	254	237	207	223	298
Female	72	76	65	95	120

These enrolments are illustrated graphically in Figure. 2.2.

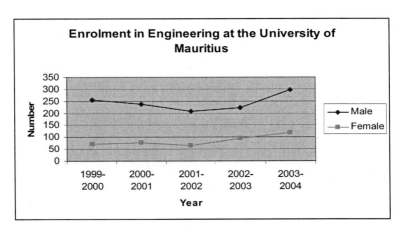

Figure 2.2: Enrolment genderwise in Engineering at the University of Mauritius

A great disparity in the enrolment of boys and girls in the Engineering field is evident, indicating perhaps a stereotype view of engineering as a male dominated area. This gender gap in participation in Engineering courses is in line with studies conducted recently and reported by Vale, Forgasz, & Horne (2004).

The second strand of research identified by Dunne and Johnson (1994) was devoted to biological factors responsible for such differences. Considerable research was conducted during the 1960's and 1970's that proposed possible biological factors as being responsible for gender differences in the learning of mathematics (Leder, 1992). Sharman noted that some of these theories have centred on a "recessive gene on the X-chromosome, the role of sex hormones, and differences in brain lateralisation" (cited in Ernest, 1994a, p. 28). On the other hand there were studies that have focused on differences in spatial visualization. Eddowes (1983) noted that boys displayed superiority in spatial and mechanical tasks in the primary years and that this gave them a better foundation for mathematics and science at the secondary school level (cited in Burton, 1986, p. 23). Also, Gallagher and De Lisi (1994) pointed out that there is a physiological difference in male and female brains and that females are better at rapid retrieval, while males are better at manipulation of information. However many studies have rejected a biological cause for gender differences in mathematics (Burton, 1986; Fennema & Sherman, 1977; Leder, 1990a; Reynolds, 2004; Tartre, 1990). It was argued that the link between spatial visualization and mathematical ability is not fully understood and that since gender differences in mathematical attainment vary across different countries, there must be other factors that affect achievement in mathematics.

The following view expressed by Mahony (1985) on biological interpretation of differences in behaviour of boys and girls is significant:

> In order to create social division between two groups some actual difference is needed as a legitimating explanatory category. Biology and difference in biology, far from explaining differences in behaviour between boys and girls, is used to give legitimacy to them. Gender differences do not flow naturally from biology but must be seen as rooted in politics. The appeal to biology is merely an excuse and as such must itself be seen as part of the rationalization of the politics of male domination.
>
> (Mahony, 1985, p. 64)

A counter example for the genetic difference was provided by Alkhateeb (2001). His study showed that high school females outperformed males in mathematics achievement. Different societal factors of Arabs in the United Arabs Emirates (high socioeconomic status, female students spending more time in indoor activities, successful adult education programs, women teachers teaching girls and men teachers teaching boys, positive images of mathematics in the Arab culture) were given to explain this phenomenon. It was argued that these findings supported the claim that gender differences in mathematics achievement are due to societal influence and not genetic. Furthermore, the analysis of the FIMS, SIMS and TIMSS have shown that gender differences in mathematics achievement varied widely from country to country and have thus questioned the validity of attributing gender differences in results to innate differences between boys and girls (Hanna, 2003).

> The IEA studies provided convincing evidence that gender differences in achievement vary widely from country to country, with the degree and direction of variation depending greatly on topic and grade level. In some countries the studies revealed marked gender differences favouring males in some topics. In other countries, no gender differences were found; and, in a few countries, the studies showed gender differences that favoured females ... in showing that gender differences in mathematics vary in magnitude and

> direction from country to country, the IEA findings call into question the
> validity of the claim made by a number of researchers that there are innate
> differences between males and females in mathematical ability. (Hanna, 2003,
> p.206)

Other studies (Sanders & Peterson, 1999) have also shown that gender differences in mathematics and science are due to social and cultural factors rather than biological factors. However, the debates on brain differences between males and females are currently being launched again (Cahill, 2005).

Other than these explanations for gender differences in mathematics, social explanations were also suspected. These constituted the third strand of research related to gender and mathematics as identified by Dunne and Johnson (1994). Several environmental and learner-related factors have been proposed in various models explaining gender differences in mathematics. In an overview of some of these models, Leder (1992) noted that:

> The various models described share a number of common features: the
> emphasis on the social environment, the influence of other significant people
> in that environment, students' reactions to the cultural and more immediate
> context in which learning takes place, the cultural and personal values
> placed on that learning and the inclusion of learner-related affective, as well
> as cognitive variables. (Leder, 1992, p. 609)

As identified by Leder (1992), the main factors that play an important role in gender differences in mathematics are:

- Environmental variables
    - School, teachers, peer group, wider society, parents.
- Learner-Related Variables
    - Cognitive Variables: Intelligence, Spatial Abilities
    - Affective Variables: Confidence, Fear of Success, Attributions, Persistence.

Leder (1992) reported that these component factors do not function in isolation, but rather that "an implicit thread running through the review is the link between the different components selected and the reciprocal interaction between factors in the environment and the individuals who function in it" (p. 610).

The environmental variables mentioned above may contribute in the reproductive cycle of Gender inequality. The following comments by McCormick (1994) give an idea of the situation.

> The assumptions underlying the stereotypes manifest themselves in different expectations for treatment of boys and girls in the classroom and this negatively affects their ability to achieve their full potential.
>
> (McCormick, 1994, p. 45).

It should be noted that if existing gender stereotypes are not challenged and the assumptions argued, they will continue to exist and be reinforced. Teachers, parents and the society at large have a major role to play in this field. The following comments from Tobias (1993) situate the cyclical nature of the issue.

> Society expects males to be better than females at mathematics. This affects attitudes; attitudes affect performance; performance affects willingness to study more mathematics; and eventually, males do better than females.
>
> (Tobias, 1993, p. 74)

Ethington(1992) proposed a psychological model of mathematics achievement which was of immense help to me in identifying the factors I wished to consider in my own research. The model is given in Figure. 2.3.

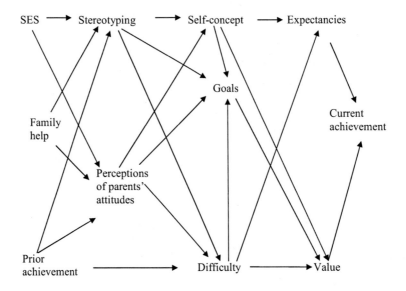

Figure 2.3: Model of achievement behavior (Ethington, 1992, p. 168)

Describing the model, Ethington(1992) noted that the first column of variables (parents' socioeconomic status, family help with the study of mathematics, and prior achievement) were referred to as 'exogenous variables' while the other variables were called 'endogenous variables' — that is, they depend on prior variables (Ethington, 1992, p. 168). She also pointed out that the paths drawn indicated the causal effects hypothesized by the model.

After having reviewed the studies which were carried over time and identified the different possible explanations that existed for gender differences in mathematics, the present study focused more on the social factors rather than biological factors in order to analyse the gender differences in the performance of boys and girls in mathematics in Mauritius. In fact, out of the factors that were identified by Leder (1992) and the model described by Ethington (1992), the focus for this present study was placed squarely on teachers, parents, peers and students' own attitude towards mathematics and the influence of these factors on the performance of boys and girls in the subject. In order to delve deeper into the issues related to these factors in my research, a further review of studies related to them was considered necessary. The following subsections review the relevant research conducted in relation to the contribution of teachers, parents, peers and students' attitude in the teaching and learning of mathematics.

Teacher variables

Teachers play a very important role in the educational setting of a child – as asserted by Hanna & Nyhof-Young (1995, p. 13):

> Teachers are one of the most important educational influences on students' learning of mathematics. The school environment or social context in which students learn mathematics is another critical factor, influencing how they learn, their expectations, their perceptions and misapprehension of mathematics and of schooling in general"
>
> (Hanna & Nyhof-Young, 1995, p.13)

The way teachers behave, interact with students and their expectations of the student's outcome and achievement have an impact on the child and the way the child will learn. Hyde and Jaffee (1998) believed that teachers and students held gender stereotypes, and that those teachers' stereotypes are activated by interactions with students. They also asserted that teachers' interactions with students affect achievement. Concerning gender difference in teacher perception, Elwood (1999) has reported that males were perceived to be more confident, positive and believed in their ability. However, girls were perceived to be less confident, more anxious and did not have enough faith in their ability and prospects for future success.

In Mauritius, teachers play a very important role in a child's development from kindergarten onwards. Their contribution to the mathematics achievement of the students too is influential and consequently an examination of the literature on this factor was seen to be important.

The impact of teacher interactions with boys and girls has been commented on by Carr, Jessup and Fuller (1999). They asserted that teacher's interactions with boys often increased their understanding and self-concept as mathematics students, while the same type of interactions lowered the self-concept of girls. As for Boaler (1997), he has shown that in a competitive environment, girls are placed in a disadvantageous position. It has also been argued that girls tended to value experiences that allowed them to think and develop their own ideas, while boys tended to emphasise speed and accuracy. Consequently this enabled boys to adapt better to the competitive environment than girls.

Communication plays an extremely important role in the teaching-learning process, and the National Council of Teachers of Mathematics has rightly placed considerable emphasis on this aspect in recent reforms in mathematics education (NCTM, 1989, 2000). Various studies have been conducted related to the issue of communication and mathematics learning (Atweh, Cooper, & Kanes, 1992; Elliot & Kenney, 1996; Stienbring, Bussi, & Sierpinska, 1998). Teacher's interactions with students in the class, the type of questions asked, reactions to responses of students, verbal and non-verbal communication strategies of the teacher all have an important influence on the teaching and learning of mathematics. The main framework in this issue is how language and other culturally symbolic systems mediate thinking. In fact, Vygotsky (1978) has argued that interactions form the social context in which children participate, and they also mediate students' thinking and learning.

In a study conducted by Khisty and Chval (2002), it was found that the teacher "plays a critical role in the communication process that forms the context of learning..." (p. 167). It has already been pointed that participation of students in classroom discussion is very crucial in a mathematics lesson. Among many other things, this allows the teacher to gain a glimpse of what is happening in the child's mind and to recognize possible misconceptions and offer remedial measures. Let us take for example a situation where the concept of "heavier" and "lighter" is being discussed. If the teacher has not varied the examples chosen and has always shown objects which are "bigger" to be "heavier", it could be that a child may develop the misconception that whenever an object is 'bigger' it has to be 'heavier'. If that student interacts with the teacher concerning this 'misconception', the teacher can immediately react by taking an example of a large piece of sponge and a small iron ball. But if the child remains silent, he or she may continue to entertain that 'misconception'.

In this review the importance of teachers in the school culture has been highlighted. How they contributed and perpetuated gender differences in their classrooms was also noted.

It should also be noted that in Mauritius, unlike in the primary sector, pre-service teacher training has not existed for the secondary level until recently. People who have obtained their academic qualifications are able to register themselves for a teaching license, and this enables them to be employed in private schools at the secondary level. To be recruited into the state schools, interviews are conducted by the Public Service Commission. It is only while they are in the service that they enroll in a pedagogical course at the Mauritius Institute of Education (some have enrolled in their pedagogical course after having taught for twenty years). Many teachers obtain advice on teaching while working and discussing with their more experienced colleagues. Full time pre-service courses have recently been introduced for graduates who wish to have a career in the teaching profession. During my career as a teacher trainer I have experienced cases where teachers used discouraging words to students who were seeking help from them, and also cases where students were negatively labeled. In this study the various interactions of the Mauritian teachers in their classrooms were studied. The differential treatment given to boys and girls as identified in the literature was examined in the Mauritian context in order to determine how it agreed or differed to those in other places of the world.

The peer influence

In the process of socialization the peer group has a fundamental role to play. The beliefs, attitudes and behaviours of individuals are to a great extent influenced by peers. In fact (Opdenakker & Van Damme, 2001, p. 408) pointed out that "The influence of peer-group background characteristics on academic achievement has been recognized for a long time". Other research studies related to the influence of peers in the learning process were conducted (DeBoer et al., 2004; Hoxby, 2002; Roscoe & Chi, n.d; Sam & Ernest, 1999), and in her research, Hoxby (2002) noted that peer effects were stronger within racial groups than between them.

The ways that groups tend to be formed are well described by Lockheed (1985)

Self-selected sex segregation is well documented as a widespread phenomenon among elementary and junior high-school aged children. It has been demonstrated in studies of student friendship choices and work partner preferences that utilize socioeconomic techniques, in surveys of student attitudes, and by observation ... Students identify same-sex but not cross-sex classmates as friends ..., choose to work with same-sex but not cross-sex classmates..., sit or work in same-sex but not cross-sex age groups..., and engage in many more same-sex than cross-sex verbal exchanges. (p. 168)

It should be noted however that this pattern has changed recently. One can observe many cross-sex groups forming in the Mauritian educational society and many exchanges occurring. The question to be asked, and one which is being addressed in this study, is whether this change in pattern has had an effect on the issue of gender and mathematics. Mauritius is a small country and one cannot really differentiate between rural and urban areas. Children move from one part of the island to other parts for school or at least for private tuition. They meet friends from different cultures and social classes and generally interact happily. However, a new phenomenon has emerged in the Mauritian society and throughout the educational world as well: violence in schools. Many cases of violence have been observed in Mauritius recently, even in primary schools. One of the important questions to ask is: Is it as a consequence of socialisation, media, or other such factors? It should be pointed out that a scientific study related to Violence in Schools in Mauritius is about to be concluded. The present study is aimed to find out the extent to which peers influence the behaviour and the learning of mathematics of their fellow students in Mauritius.

Parental support

Parents play an important role in influencing children's attitude towards mathematics (Ethington, 1992). It is argued that parents' education and occupation play an important role in women's performance in academic matters (Oakes, 1990). It was also asserted that "parents gender-stereotyped beliefs are a cause of sex differences in students' attitudes toward mathematics"(Eccles & Jacobs, 1986, p.375). The contribution of parents to gender differences in mathematics achievement has been the subject of considerable research. For instance, Lummis & Stevenson (1990, cited in Gutbezahl, 1995, p.1) stated that "By the time children enter kindergarten, parents expect girls to do better at verbal tasks and boys to do better at math." These negative expectations of parents tend to put girls at a disadvantageous position in mathematics classroom. Even concerning toys, parents have different treatment for boys and girls: Boys are given more spatially complex toys and are also given more opportunities to explore their physical worlds. These factors may contribute to the gender differences in spatial ability which is an extremely important component of mathematics. Concerning the relationship between parental support and mathematics achievement, Telfer & Lupart (2001, p.9) have reported that "children who have relationships with their parents that are free of anger and disappointment will achieve better in math and science."

Research has also shown that

> ...lower parent educational experiences – especially for mothers – is a significant predictor of how and when students will receive special education. In fact, children whose mothers completed college will receive special education over two years earlier in their educational profile than those whose mothers only finished the eighth grade" (George, 1996, cited in Hammrich, 2002, para 7).

Ma (1999) has proposed a theoretical explanation of how parental involvement helps children in their studies. He argues that involvement of parents improves children's cognitive skills and this consequently helps them to succeed in academic tasks. In a recent study, Raty, Vanska, Kasanen, & Karkkainen (2002, p. 126) noted the following concerning 'differential parents' attribution' to the mathematical success of boys and girls:

> The parents accounts of their child's success in mathematics contained 'the gender-related attribution', that is the parents rated talent as a more important reason for their child's mathematical success than did the parents of girls. In contrast, the parents of girls rated effort as a more important reason for their child's mathematical success than did the parents of boys.

After having reviewed the contribution of parents to the issue of gender and mathematics internationally, the question to be asked is: How does it relate to the situation in a developing country? Mauritius has undergone much development since independence. At first the Mauritian economy was based principally on sugar and considerable developments took place during the Agricultural revolution. This was then followed by Industrial revolution where textile industries became a major pillar of the Mauritian economy. The Tourism industry also flourished more recently to provide another support to the Mauritian economy. These developments called for more involvement of the population of Mauritius, especially those of the working class. The implication of these developments in the education of children in Mauritius is worth determining. Do the students in a developing country such as Mauritius receive sufficient parental attention and support in their education? Does this support vary amongst different social classes in Mauritius?; Among different ethnic communities? What are the parental views of mathematics in Mauritius? This study aimed to find some answers to these pertinent questions.

Attitude towards mathematics

The role that the affective component plays in gender-related differences in mathematics achievement has been the subject of considerable research (Ma & Kishore, 1997; Seegers & Boekaerts, 1996). Aronson, Wilson & Akert (1997, p. 229) define attitude as:

> ... attitudes are made up of an affective component, consisting of your emotional reactions toward the attitude object (e.g. another person or a social issue); a cognitive component, consisting of your thoughts and beliefs about the attitude object; and a behavioural component, consisting of your actions or observable behaviour toward the attitude object.

(Aronson et al., 1997, p. 229)

Several studies have shown that interests influence academic achievement and learning in school (Krapp, 1998a, 1998b, 1999; Schiefele, Krapp & Winteler, 1992; cited in Köller et al., 2001). On the other hand, in a recent study, Ma and Xu (2004, p. 275) have shown that "the causal ordering between attitude toward mathematics and achievement in mathematics is predominantly unidirectional from achievement to attitude across the entire secondary school".

Reasons for girls' negative attitudes towards mathematics and for being more negative towards mathematics than boys have been suggested by Dweck (1986; cited in Gutbezahl, 1995). According to his study, girls tend to hold an entity theory of intelligence; that is intelligence is something that a person has or does not have. However, boys tend to hold an incremental theory of intelligence; that is intelligence in a given domain may be increased by hard work. Combined with the differential treatment in the mathematics classroom and the entity theory of intelligence, girls tend to attribute their failure in mathematics to lack of ability and consequently give up trying hard.

In this present study, the attitude towards mathematics of a representative sample of the students in Mauritius was determined using the Modified Fennema-Sherman Mathematics Attitude Scale questionnaire. It consists of 47 items dealing with four subscales: Confidence, Usefulness, Mathematics as a Male Domain and Teacher Perception. Further issues concerning the students' attitude towards mathematics were to be probed in through interviews. A literature search was considered necessary to frame questions to be asked during the students' interviews.

Many researchers have underlined the relationship between student confidence in their mathematical ability and their achievement in the subject. Confidence "is one part of self concept and has to do with how sure a student is of his or her ability to learn new mathematics and to do well on mathematical tasks. Confidence influences a student's willingness to approach new material and to persist when the materials become difficult" (Meyer & Koehler, 1990). Confidence plays an important role with students deciding to continue or discontinue studying mathematics (Leder, Forgasz, & Solar, 1996). Armstrong and Price (1982) found confidence in learning mathematics to be the second most important variable affecting students' participation in mathematics for females and the third most important for males. In another study, Wolleat, Pedro, Becker and Fennema (1980) asserted that "females, as compared to males, have been found to be less confident about their ability to learn mathematics, to underestimate their ability to solve mathematical problems, and to believe to a lesser degree that mathematics will be personally useful" (p. 356)

Researchers have also been interested in the ways students attribute the cause of success or failure in mathematics. Weiner's Attribution Theory (1971) played an important role in this venture. Weiner identified ability, effort, task difficulty and luck as the most likely causes attributed to various academic outcomes. The dimensions of stability and locus of causality were attributed to each of these causes. Stability of the perceived cause refers to the endurance of the particular attribution. Since ability and objective task difficulty are enduring, they are termed as stable. On the other hand, effort and luck are transient and thus unstable. Locus of causality refers to the origin of the perceived reason of the outcome. Ability and effort are internal to the individual, while luck and task difficulty are external. The reason attributed to success or failure plays an important in subsequent occurrence of success or failure. If an outcome is attributed to factors which are stable, then that outcome is more likely to occur in the future. Reasons that are normally attributed to success and failure are ability, effort, task difficulty and luck (Weiner, 1974, cited in Fullarton, 1993). Research findings (Leder, 1992) have established that females attributed success and failure in mathematics more strongly than males in accordance with a pattern described as 'learned helplessness'. Ernest (1994b) described the way boys/girls tend to attribute success or failure in mathematics.

	Success at mathematics	**Lack of success**
Boys	Skill, ability	Bad luck, lack of effort
Girls	Good luck, effort	Lack of skill and ability

The same conclusions were reached by Leder, Forgasz, and Solar (1996).

This study aimed to determine the level of confidence of the boys and girls in Mauritius in learning mathematics, and the factors to which they attribute their success/failure in this subject.

The educational system of Mauritius has a peculiar characteristic which differentiates it to a great extent from the developed countries: the practice of private tuition. The Government of Mauritius has shown some concern over this problem and has legislated to declare illegal the practice of private tuition at the lower primary level (average age 6 to 8 years). However, at the upper primary level and upper secondary level, private tuition is legal and very popular. Mathematics is a subject where the majority of students taking part in the School Certificate and Higher School Certificate examinations take private tuition. There are even cases where students take private tuition in mathematics from two different teachers (other than the usual classroom teacher). It was worth analysing the point of view of students, teachers and parents to this phenomenon and what they feel is its contribution in the teaching and learning of mathematics at the secondary level. Such an investigation would throw light on some issues that have not been examined in depth previously in Mauritius.

The Learning Environment

My first experience with Learning Environment research and using the tools available to investigate students' perception of various aspects of the classroom learning environment occured when I was enrolled in a unit of coursework for my doctoral programme. I was amazed with the variety of available valid questionnaires designed for that purpose. Woolfolk (1998) described the classroom as an ecological system and stated that "the environment of the classroom and the inhabitants of that environment — students and teachers — are constantly interacting (p. 440). Each partner in the educational family should contribute towards creating and maintaining a positive learning environment for the child. This section presents a brief review of research conducted in that field which guided me towards the learning environment tools to use in my study, what analysis to perform and how the results can be interpreted.

The importance of assessing the learning environment in the teaching and learning process has been emphasized in many research studies (Fraser, 1998). It has been noted that students' perceptions of their classroom psychosocial environment have a significant influence on their cognitive and affective learning outcomes (McRobbie & Fraser, 1993; Yarrow, Millwater, & Fraser, 1997). Research on learning environment can be traced to Lewin's classic human behaviour definition represented by B= f (P, E); that is behaviour (B) is a function of the person (P) and his environment (E) (Fraser, 1998). Over 35 years ago, Walberg (1968) developed the instrument *Learning Environment Inventory* (LEI) to assess the learning environment in physics classrooms. Independently Moos and Trickett (1974) developed a series of environment measures which concluded with the Classroom Environment Scale (CES) asking students for their perception of the learning environment of the class as a whole.

For assessing any human environment, Moos (1979) defined three basic types of dimensions:

(a) Relationship dimensions - which identify the nature and intensity of personal relationships within the environment and assess the extent to which people are involved in the environment and support and help each other.

(b) Personal development dimensions - which assess basic directions along which personal growth and self-enhancement tend to occur.

(c) System maintenance and system change dimensions - which involve the extent to which the environment is orderly, clear in expectations, maintains control, and is responsive to change.

These two instruments can be considered to be the driving forces for the study of classroom learning environments and development of other instruments for such evaluation. Indeed, many economical, valid and relevant questionnaires are available for assessing students' perceptions of the classroom environment (Fraser, 1998). It should be mentioned that appropriate instruments have been designed to evaluate learning environment from primary to tertiary levels. A summary of some of these instruments is given in Table 2.4.

In Mauritius the field of learning environment in educational research is a recent one, with very few studies conducted. In one study conducted by Bessoondyal and Fisher (2003) the perceptions of preservice primary school teacher trainees concerning teacher-students interactions in a mathematics classroom in the teacher training institution was carried out, and it was found that the classroom environment concerning teacher interaction was positive; the mean score being quite high for the Leadership, Understanding, and Helping/Friendly scales, and the mean score quite low for the Uncertain, Admonishing and Dissatisfied scales.

As the present study is classroom-focused, an evaluation of the type of interactions in the secondary school mathematics classes in Mauritius was considered important in order that remedial measures could be introduced in the teaching and learning package that was to be designed. Consequently, the Questionnaire on Teacher Interaction (QTI) was chosen to be used in Phase two of the study. After the implementation phase, where certain strategies were put forward, an evaluation of students' views related to their involvement in tasks, the extent of cooperation in the class and equity proved to be important. The "What Is Happening In this Class?" (WIHIC) questionnaire was chosen to be used in Phase three. These two questionnaires will be described in later chapters and different studies related to their use will be discussed also.

Table 2.4: Overview of some Classroom Environment Instruments

Instrument	Level	Items per scale	Relationship dimensions	Personal development dimensions	System maintenance and change dimensions
Learning Environment Inventory (LEI)	Secondary	7	Cohesiveness Friction Favouritism Cliqueness Satisfaction Apathy	Speed Difficulty Competitiveness	Diversity Formality Material environment Goal direction Disorganisation Democracy
Classroom Environment Scale (CES)	Secondary	10	Involvement Affiliation Teacher support	Task orientation Competition	Order and organisation Rule clarity Teacher control Innovation
Individualised Classroom Environment Questionnaire (ICEQ)	Secondary	10	Personalisation Participation	Independence Investigation	Differentiation
My Class Inventory (MCI)	Elementary	6 - 9	Cohesiveness Friction Satisfaction	Difficulty Competitiveness	
College and University Classroom Environment Inventory (CUCEI)	Higher education	7	Personalisation Involvement Student cohesiveness Satisfaction	Task orientation	Innovation Individualisation
Questionnaire on Teacher Interaction (QTI)	Secondary / Primary	8- 10	Helpful/ friendly Understanding Dissatisfied Admonishing		Leadership Student responsibility and freedom Uncertain Strict
Science Laboratory Environment Inventory (SLEI)	Upper Secondary/ Higher education	7	Student cohesiveness	Open-Endedness Integration	Rule clarity Material environment
Constructivist Learning Environment Survey (CLES)	Secondary	7	Personal relevance Uncertainty	Critical voice Shared control	Student negotiation
What Is Happening In this Class? (WIHIC)	Secondary	8	Student cohesiveness Teacher support Involvement	Investigation Task orientation Cooperation	Equity

The reform movement in teaching mathematics for understanding

The need for effective teaching of mathematics in the classroom has been made several times and in many parts of the world. Many forums and studies have been conducted to discuss this important issue (Grouws, Cooney, & Jones, 1989; Jaworski, 1994, 2003; Malone & Taylor, 1993; NCTM, 2000; Nickson, 2000; Treagust, Duit, & Fraser, 1996).

The following comments from the Principles and Standards for School Mathematics (NCTM, 2000, p.16-17) outline the importance of effective teaching:

> Students learn mathematics through the experiences that teachers provide. Thus, students' understandings of mathematics, their ability to use it to solve problems, and their confidence in, and disposition toward mathematics are all shaped by the teaching they encounter in school. The improvement of mathematics education for all students requires effective mathematics teaching in all classrooms."

Reforms in mathematics education call from a shift from instruction that encourages memorization of definitions and use of rules and procedures toward instruction that emphasizes mathematical inquiry and conceptual understanding (Saxe, Gearhart, & Nasir, 2001).

The term "teaching for meaning" has been used by Knapp, Shields & Turnbull (1995) to refer to the following alternatives to conventional practice:

1. instruction that helps students perceive the relationship of 'parts' to wholes

2. instruction that provides students with the tools to construct meaning in their encounters with academic tasks and in the world in which they live, and

3. instruction that makes explicit connections between one subject area and the next and between what is learned at school and children's home lives.

In their study, Knapp et al. (1995) recorded evidence that students who were exposed to instruction, emphasizing meaning, performed significantly better on advanced mathematical skills than those students who were exposed to conventional sessions. They further reported that this significant difference in achievement was noted both for high and low performing children and in schools in deprived areas.

Another concern in the mathematics educators' community is the inability of students to use mathematics learnt at school in situations outside the classroom context (Boaler, 1998). This lack of transfer of the mathematics learnt in schools to other situations has been argued to be due to the fact that students did not fully understand the concepts (Boaler, 1998). Boaler (1998) argued that enough opportunities should be given to students to deal with real-life mathematical situations within the school context. This exposure tends to facilitate the use of appropriate mathematical knowledge to deal with similar tasks in the real world.

Boaler (1998) strongly advocated appropriate teaching strategies that would help students perform the smooth transition of their mathematical knowledge to novel situations in the real world. In this study, Boaler concluded that, if students, based on the experiences encountered in classrooms, view mathematics as a set of rules and procedures that should be remembered and followed to solve problems, these students tended to encounter problems when the situations they were faced to were slightly different from what they are normally used to. On the other hand, students who believed that mathematics involved active and flexible thought and who had developed an ability to adapt and change methods to fit new situations, tended to deal with situations in the real world more successfully.

Motivation has been linked to achievement and much research has been carried out in this field (Beghetto, 2004; Ethington, 1992; Leder et al., 2002; Peterson, 1988; Pintrich & Schunk, 2002; Schiefele & Csikszentmihalyi, 1995; Stevens, Olivarez Jr, & Hamman, 2005).

Motivation is usually defined as an internal state that arouses, directs, and maintains behaviour (Woolfolk, 1998). It has been described as having three psychological functions:

- energizing or activating behaviour – what gets students engaged in or turned off learning
- directing behaviour – why one course of action is chosen over another
- regulating persistence of behaviour – why students persist towards goals.

(Ford, 1992, cited in Alderman, 1999)

Motivation propels and encourages students to be fully engaged in academic activities. Motivated students strive to understand the underlying concepts and tend to use higher cognitive processes to learn. Motivation to learn can be intrinsic or extrinsic. Intrinsic motivation has been described as the drive or desire of the student to engage in learning 'for its own sake' (Middleton & Spanias, 1999, p.66). Students who are intrinsically motivated engage in academic tasks because they enjoy them. Students who are extrinsically motivated engage in academic tasks to obtain rewards (such as teacher's praise, approval of their participation in a lesson) or to avoid punishment (such as poor grades, disapproval). Recent research (Cameron, Pierce, Banko, & Gear, 2005, p. 654) has shown that

> ...giving rewards for successful achievement on an activity leads individuals to express high task interest and to be motivated to perform the activity and similar other tasks in the future. The study also shows that rewards can activate processes that involve both internal and external sources of motivation.

Classroom environment plays an important role in students' motivation, which in turn influences achievement. Bong (2005) showed that changes in students' perception of their learning environment brought about changes in their personal motivation. In another study, Schiefele & Csikszentmihalyi (1995) noted the significant relationship between the students' experiences in the mathematics classroom and their interest, which often predicted achievement in mathematics. Other studies have shown that increased intrinsic motivation is related to greater conceptual understanding of mathematics (Nichols, 1996; Stipek et. al., 1998; Valas & Slovik, 1993; cited in Stevens, 2005).

One of the fundamental assumptions for the third phase of this project was based upon my belief that an improved classroom environment, where teaching is student-centered, where students are involved in their own construction of knowledge through well designed and relevant activities, and which involve the proper use of cooperative learning, does enhance the intrinsic motivation of the students to learn mathematics. This should eventually result in an enhancement of the students' performance and achievement in the subject and also an improvement in their attitude towards mathematics.

Understanding is described in terms of the way an individual's internal representations are structured. Hiebert & Carpenter (1992, p. 67) highlighted that "A mathematical idea, procedure or fact is understood thoroughly if it is linked to existing networks with stronger or more numerous connections". It is claimed (Hiebert & Carpenter, 1992) that students who understand the mathematical concepts will retain what they learn and transfer it to novel situations. In fact the researchers noted the following in relation to understanding:

- Understanding is generative
- Understanding promotes remembering
- Understanding reduces the amount that must be remembered
- Understanding enhances transfer
- Understanding influences beliefs

The need for context in problems to assess students' mathematical understanding was highlighted by Van den Heuvel- Pahhuizen (2005). It was mentioned that contexts help in

- enhancing the accessibility to problems
- contributing to the transparency and elasticity of problems, and
- suggesting solution strategies to students.

Van den Heuvel-Panhuizen (2005, p. 2)

The importance of contexts in assessing students' mathematical understanding was further supported by other studies (Kastberg, D'Ambrosio, McDermott, & Saada, 2005).

To cater for higher-order learning, Peterson (1989) advocated a focus on the meaning of learning tasks, the involvement of students, and the teaching of higher-level cognitive processes and strategies. The constructivist theory of learning is one that might achieve this goal. The constructivist view was a "powerful driving force in science and mathematics education, particularly during the past decade" (Treagust et al., 1996, p. 3). Several studies related to the teaching and learning of mathematics using a constructivist paradigm was reported in Malone & Taylor (1993). The constructivist perspective asserts that conceptual knowledge cannot be transferred, carefully packaged, from one person to another. Rather, it must be constructed by each child on the basis of her own experience. Children should be provided with activities that are likely to generate genuine mathematical problems, ones that can provide them with opportunities to reflect and reorganize their existing ways of thinking. Nickson (2000) outlines three major implications of a constructivist approach for the effective teaching and learning of mathematics: (1) children are makers of knowledge rather than receivers of knowledge, (2) children need to experience mathematics in a context, and (3) each child's contribution in a mathematics lesson need to be acknowledged and considered.

Convinced by these ideas, I devised the teaching and learning package for the third phase of this present study using a constructivist perspective. Lessons were devised in such a way that students were given opportunities to be actively involved in tasks, construct their own knowledge with necessary scaffolding where needed. While designing the lessons, particular attention was paid to the following issues:

1. The cognitive complexity of the mathematical concept, procedure, or process involved,
2. The mental processes that children were expected to develop,
3. The prerequisite knowledge and experiences children must have to be involved in the tasks, and
4. The nature of the learning experiences that might facilitate children's mathematical development.

A detailed discussion of this package is provided in Chapter Six.

Language and mathematics

Language plays an important role in the teaching and learning of mathematics (Barwell, 2005; Clarkson, 1992; Clarkson & Galbraith, 1992; Cuevas, 1984; Ellerton & Clarkson, 1996; Kaphesi, 2003; Laborde, 1990). The teaching and learning of mathematics require various language activities such as reading, listening, writing and discussing (Laborde, 1990). But these activities were found to be difficult to achieve when the language through which the students have access to the mathematical content is a bar to both understanding and expression (Laborde, 1990).

In another study, Ellerton et al. (1996, p. 1001) reported that "far more children experience difficulty with the semantic structures, the vocabulary, and the symbolism of mathematics than with standard algorithms". They further argued that "semantic structure has a much more important influence on learning and the quality of participation in classroom discourse than other more obvious language variables (such as vocabulary)"(p. 1004).

For learners who speak two or more languages, it was found that the interplay in the learning process between the language codes may either assist or detract learning (Ellerton & Clarkson, 1996). Learners may have a cognitive advantage if a threshold of competence in the two (or more) languages has been reached. However, if learners are expected to use only the language of instruction, then negative outcomes may follow. In a recent study conducted in Malawi, Kaphesi (2003, p. 265) noted that "many problems that learners of mathematics encounter are partly due to the inability to cope with the demands of the language of instruction". Along the same lines, Morgan (1999) pointed out that teachers should be aware of the problems encountered by students while they struggle to cope with the discussion in the classroom and reading and understanding of written materials. Students need assistance and encouragement to get involved in a proper classroom discussion.

Concerning difficulties related to solving word problems, Laborde (1990) highlighted that

> Understanding what is to be solved requires understanding the problem statement given in an oral of written form.... The wording of the problems appears to influence the students' problem representations and therefore their strategies of solution
>
> Laborde (1990, p. 62)

It has been reported that teachers, especially those of Limited English Proficient (ELP) students, may be tempted to use key words as "signaling" certain operations. It should be noted that this practice may lead to misconceptions that may hamper understanding of mathematical concepts. The Newman procedure to deal with a written mathematical task has been described (Ellerton & Clarkson, 1996) as follows: Reading (or Decoding), Comprehension, Transformation (or Mathematising), Process Skills, and Encoding. Ellerton et al. (1996, p. 1001) reported that "far more children experience difficulty with the semantic structures, the vocabulary, and the symbolism of mathematics than with standard algorithms". They further argued that "semantic structure has a much more important influence on learning and the quality of participation in classroom discourse than other more obvious language variables (such as vocabulary)"(p. 1004). In a recent study, Barwell (2005)noted that

> ...the participants in the study used attention to narrative experience to negotiate a shared understanding of a word problem, to relate the context of the word problem to their own experience, and to negotiate their relationships with each other. These findings suggest that opportunities for students to draw on their experience contribute to a supportive linguistic context within which they are able to work on their mathematical task

> (Barwell, 2005, pp. 345-346).

The issue of language and mathematics is of vital importance in the Mauritian context as Mauritius is a multilingual society where the medium of instruction is English, the language most often spoken is French, together with an L_1 which is Creole. This is the language used most often at home. There do exist other Asian languages like Hindi, Urdu, Tamil, Telegu, Marathi, and Chinese which are studied mostly officially at school. It should be mentioned that right from the primary schooling (average age of five years) the Mauritian children are exposed to at least three languages; English and French as subjects and English being the medium of instruction and Creole as the most common spoken language at home. Many students are exposed to another language at school called Oriental Language (Hindi. Urdu, Tamil, Telegu, Marathi, or Chinese). It should be mentioned that very often French is used as a support language in the teaching process at the primary level. It is worth finding out the extent to which the students in Mauritius are coping with this important issue of language in their learning of mathematics.

Conceptual framework for this study: Equity in education

Gender equity in mathematics has been the focus of many studies (Carey, Fennema, Carpenter, & Franke, 1995; Fennema, 2000; Frankenstein, 1995; Goodell & Parker, 2001; Goodell & Parker, 2003; Hanna, 1996; Leder, 1995, 1996; Silver, Smith, & Nelson, 1995; Stevens et al., 2005). Access to basic education and the promotion of equity and social justice have been the concern of various international educational organizations. Different forums have been set up where the emphasis was placed on education as a right and an asset that should be made available to all. Indeed in the *World Conference on Education for All* held in Jomtien in 1990, a declaration was adopted which stated that "The most urgent priority is to ensure access to education of girls and women and to remove every obstacle that hinders their active participation" (UNESCO, 1990, p.9). Many countries have adopted policies of equality of access to education as recommended by international forums, but merely having laws relating to equality of access does not in itself ascertain equity in education. In Mauritius, Bunwaree (2002, p.4) pointed out that "it is not sufficient to simply provide access to schools, but it is the outcomes of schooling and the life chances that schooling contributes to which are important". Different measures are being taken by the government of Mauritius in the field of education. In fact, in its effort to offer more opportunities for students in Mauritius to have access to tertiary education, the Government of Mauritius has recently allowed more foreign universities to offer their courses to Mauritians. The government is also opening another university in the country to be named the Open University of Mauritius. The main objective of this university will be to offer courses on a distance education mode.

As the overall aim of this study was to devise ways and means to help boys and girls in Mauritius learn mathematics in a more meaningful way, ascertaining gender equity as far as mathematics education is concerned was of interest. The document published by the American Association of University Women Educational Foundation (1998, p. 2) highlighted that:

> When equity is the goal, all gaps in performance warrant attention, regardless of whether they disadvantage boys or girls. Rather than hold girls to boys standards, or vice versa, schools need to give students the resources each needs

to achieve a universally held high standard. In a gender-equitable and rigorous school system, gender gaps would be insignificant and all students would excel.

Many factors affecting enrolment and achievement in mathematics and mathematics-related courses were noted and concern about gender equity raised. One of the ways of analyzing the problems was to probe deep inside the schools and classrooms, as highlighted in the *World Declaration on Education for All* (UNESCO, 1990):

> … the main focus of basic education must be on actual learning acquisition and outcome rather than exclusively upon enrolment, continued participation in organized programmes and completion of certificate requirements. (UNESCO, 1990, p.9).

Indeed much research has been carried out internationally and different intervention programs were designed to help in achieving gender equity. It has been found that in many countries gender differences have diminished tremendously, as asserted by Hanna (2003, p. 209):

> Gender differences in mathematics achievement at age 13 have decreased dramatically and almost disappeared in all the participating countries (of the TIMMS). In effect, gender equity has been reached for this age group. At age 17, on the other hand, boys are still doing better than girls in some areas of mathematics, though the gender gap has considerably decreased over the years 1964 to 1995.

It is clear, however, that full gender equity has not yet been achieved. There is still a significant difference in the participation of men and women in job markets, especially in scientific fields.

In terms of representation, full gender equity has not been reached, despite numerous policies and legal measures put in place to encourage it. Women have achieved a considerable presence at all levels of education over the past few decades and indeed have made a substantial advance in the sciences. In certain scientific disciplines, however, notably mathematics, physics, and engineering, their presence still lags behind that of men.

(Hanna, 2003, p.210)

In Mauritius much effort has, and still is, being expended by the government and educational authorities to bring an equitable education to the people of the island. With a view to offer wider educational opportunities, secondary education was made free in 1977. Over the years it has been found that, despite this policy of free education, parents from the low-income groups are encountering more and more problems in educating their children. The private expenditure on education has been increasing and some parents continue to find it difficult to cope. Consequently, many programs have been put in place by the government to ease the financial burden. These programs include the distribution of free textbooks and assistance for paying examination fees. It should be pointed out that parents who are eligible for these benefits have to satisfy certain criteria laid down by the Ministry of Social Services. It should also be highlighted that private tuition is almost a parallel institution to the schooling system, at least for the upper grades of the secondary level (from Form IV). Private tuition adds to the financial burden of parents.

The position of the Mauritian Education in the African region noted in a draft document entitled *Mauritius Education and Training Sector* (2003) is as follows: "Although Mauritius is one of the few African countries which has nearly achieved universal primary education, only around 65% of students make transition into secondary level and less than half graduate from 6th form colleges" (Chapter five, p. 3). It can be found that despite the official policy of free access for all, the realities are different. Thirty-five percent of the young population of Mauritius leaves the academic field after only six years. The pyramidal structure of the education system is being questioned and different committees have been set up by the Ministry of Education and Scientific Research to propose alternative systems to care for all the children of Mauritius. One measure which has already taken by the Mauritian government is to make secondary education compulsory up to the age of 16 as from 2005. Different prevocational streams have been created, and a programme to cater for further education is being developed presently by the Ministry.

An analysis of the results of the performance of boys and girls in mathematics at the School Certificate level (Chapter One) did reveal a significant gender difference. In view of the place and position that mathematics enjoys in the Mauritian society, for further education or job markets, these differences tend to create further inequities in the society. For comparison purposes, the numbers of male and female academic staff in the department of mathematics in the three tertiary institutions are given in Table 2.5.

Table 2.5: Academic staff in the mathematics department

	University of Mauritius	Mauritius Institute of Education	University of Technology
Male	8	6	5
Female	3	0	1

The three departments are obviously heavily male biased.

The contribution of women in the economic, social and educational developments of our country is worth noting also. At the time of agricultural revolution women worked in sugar cane fields alongside of men, after having ensured that household obligations had been met. They contributed under different roles: looking after household matters, income earner and the rearing of the children. The main responsibility of child rearing was on the mother, and they were responsible for the children's informal and cultural education. As the other revolutions occurred, more women were involved in the labour market, in offices and textile factories. Another major change that was noted was the shift from an extended family system to nuclear family system. This brought a change in the way of life and the informal education of the children. Today one can find women involved in many high positions in the society. At a recent meeting held in August 2004 concerning women and decision making, Mrs. Prabha Chinien, Registrar of Companies asserted that she never felt that being a women could ever pose a problem as far as her professional life was concerned (Savripene, 2004). She further added that this conviction was due to her parents' education and policy for equality of opportunity. Other speakers pointed out that much effort has to be made to reconcile their professional and family obligations. Nowadays more and more women are involved in the politics. Discussions are presently under way for changes to be brought into the legislative system to ascertain that a certain quota of seats in the Assembly be reserved for women. An interesting question to ask is: In what ways has this change affected the education of Mauritian children?

It should also be pointed out that the quality of results of Mauritian students in general is far from being acceptable. Around 50% of the students taking the School Certificate examination fail to score a 'credit' in mathematics. There seems to be a definite problem as these students have been studying mathematics already for 11 years. Furthermore, since 1999, the scripts of School Certificate examination Paper 1, and since 2002 the scripts of Paper 2 are corrected in Mauritius. This has given the local teachers the opportunities of noting the kind of mistakes certain children make in mathematics. At times they seem to be astonished as all the students who take part in the School Certificate examination have passed the CPE examination at the end of primary level and thus have a basic notion of the concepts in mathematics. It has been reported that there are students who cannot solve the problems they must have been solving some five or six years earlier, thus there appears to be 'degraduation' in the learning of mathematics among these students. This is definitely a cause of concern and strategies need to be found out to help all the boys and girls in Mauritius to improve in their mathematics achievement at the secondary level.

In this study, the difficulties that our Mauritian students are encountering in the learning of mathematics were analysed. The impact of the different factors and reasons identified in the literature review in the Mauritian context were examined. One finds that in the Western and other developed countries, gender difference in mathematics achievement have decreased considerably and are in certain cases insignificant. This may be as a result of the consciousness raised on these issues and the different intervention programs implemented to that effect. The main aim of this study was to analyse this problem in a developing country which gained independence 36 years ago. What measures can be taken to ensure gender equity in mathematics classrooms which could help in achieving the goal set by the Mauritian Government of becoming a Cyber Island in the Indian Ocean and African Region? Answers to these questions may provide other developing countries with some ways and means to deal with the issue of gender equity in mathematics and suggest ways to enhance the teaching and learning of mathematics at the secondary level.

Summary of the chapter

In this chapter different studies related to gender and mathematics within a chronological perspective were reviewed. The different factors that tend to affect the achievement of boys and girls in mathematics were analysed. How different intervention programs and strategies were developed and implemented to deal with the issue of gender differences in mathematics was also noted. The varied degrees of success of these ways to achieve gender equity in mathematics education in many countries, ranging from almost eradicating gender differences at the age of 13 to diminishing gender differences at the age of 17, was also found. Cases where gender differences still exist have also been noted. A review of the policy regarding gender equality in education in Mauritius has been presented together with some of the underlying realities. It has been found that much effort is being expended by the government in that direction, but still more is required. An attempt has also been made to indicate where the study fits into the literature base and how it hopes to widen the base.

The next chapter describes the research design, methods used and ethical issues taken into consideration in this study.

CHAPTER THREE
Methodology

This chapter describes the methodology chosen and used for the study. The main aim of the study was to identify factors that impact on boys' and girls' mathematics achievement at the secondary school level in Mauritius and devise strategies to enhance their achievement in mathematics and attitude towards the subject. Specifically, it investigated the types of difficulties that boys and girls in Mauritius encounter while learning mathematics. It was explained in Chapter One that at the end of primary level national examinations there is no significant gender difference concerning the performance of students in mathematics in Mauritius. However, by the time of the School Certificate examinations (average age 16 years), significant gender differences in mathematics can be noted (especially in terms of quality of performance). This is in line with research findings as stipulated by Fennema (2000, para. 10):

> ...there were differences between girls' and boys' learning of mathematics, particularly in activities that required complex reasoning that the differences increased at about the onset of adolescence...

There is an urgent need to analyse the reasons which have brought this change in performance. The issues to be addressed are: what factors have contributed to this significant change in the mathematics performance of boys and girls in Mauritius at the secondary level? To what extent is the hierarchical nature of mathematics accounting for such differences? What are the roles of teachers, parents and peers, as well as students' attitude towards mathematics in determining the students' achievement in mathematics at the secondary level? The research questions of this present study are as follows:

1. What are the factors that contribute to the mathematical achievement of Mauritian students at the secondary school level?

2. What types of difficulties do Mauritian boys and girls encounter while learning mathematics at the secondary level?

3. Why do these difficulties occur?

4. How effective is a teaching and learning package based on the findings of (1) above at enhancing the attitudes and mathematical achievement of secondary school students?

To address these fundamental issues, three phases of the study were developed. For the first phase a survey method was used and involved 17 schools across the island (including Rodrigues) in order to obtain a general idea of the performance of boys and girls across Mauritius. The way the schools were selected is discussed in more details in another section of this chapter. I visited the schools in person and explain the objectives of the study to the rector of the school and the head of mathematics department. One class of Form IV (15 years on average) was selected from each school and the students were administered a questionnaire which consisted of some background information and items on the mathematics curriculum. Students were expected to work out these questions which were then corrected and analysed by me. A questionnaire (see Appendix Three) dealing with attitude towards mathematics was also administered to these students. Different statistical analyses were carried out to examine the response of boys and girls to the different items.

After gaining an overall picture of the situation across Mauritius, it was necessary to perform a more in-depth analysis. Consequently, a case-study approach was adopted for the second phase of the study. The importance and relevance of qualitative research has been highlighted by Denzin & Lincoln (Denzin & Lincoln, 2000):

> Qualitative research involves the studied use and collection of a variety of empirical materials – case study; personal experience; introspection; life story; interview; artifacts; cultural texts and productions; observational, historical, interactional, and visual texts – that describe routine and problematic moments and meanings in individual' lives. Accordingly, qualitative research deploy a wide range of interconnected interpretive practices, hoping always to get a better understanding of the subject matter at hand.
>
> (Denzin & Lincoln, 2000, pp.3-4)

Four secondary schools were selected (one single boys', one single girls' and two coeducational schools) for this purpose. A class of Form IV was identified in each of the schools with the help of the rector and the Head of Mathematics department. Classroom observations were carried out in these classes for a period of four months following which the mathematics questionnaire used in the first phase was administered to the students in the sample. Interviews of students, teachers, rectors and parents also were conducted, keeping in mind the main research questions of the study.

The study was conducted using a mixed mode, as recommended in the literature. "Combining methods may be done for supplementary, complementary, informational, developmental, and other reasons (Strauss & Corbin, 1998, p.28). The essence of using both approaches has also been highlighted by Breitmayer, Ayers, and Knafl, 1993. "Each adds something essential to the ultimate findings, even to the final theory…" (cited in Strauss & Corbin, 1998, p.28). There can be back-and-forth interplay between combinations of both types of procedures, with qualitative data affecting quantitative analyses and vice versa. The issue in this study was mainly concerned with how the combination of methods could work together to foster the development of theory.

The present chapter has been structured so that the research design of the study is discussed first. This is then followed by an explanation of how the samples for the different phases of the study were chosen. A detailed description of the three phases is then provided.

The methodological framework of the study: The interpretive model of research

An interpretive model of research was chosen for this study. Interpretivism is often described as having a contrasting epistemology to positivism. The distinction between the two is highlighted by Cohen, Manion and Morrison (2000, pp. 27-28).

> Positivist and interpretive paradigms are essentially concerned with understanding phenomena through two different lenses. Positivism strives for objectivity, measurability, predictability, controllability, patterning, the construction of laws and rules of behaviour, and the ascription of causality; the interpretive paradigms strive to understand and interpret the world in terms of its actors. In the former, observed phenomena are important; in the latter meanings and interpretations are paramount.

It should be noted that the interpretive paradigm is characterized by a concern for the individual and its main endeavour is to understand the subjective world of human experience. Indeed, in this study, interpretations concerning Mauritian students' responses to the teaching and learning of mathematics were the main concern. The realities of their interaction with teachers, parents, peers and themselves and their views about mathematics in general were the factors that had to be determined.

Ontological assumptions

One of the main reasons for choosing the interpretive model of research for this study was because education and knowledge are socially and culturally bound together and these aspects have to be taken into consideration when a study on teaching and learning is conducted. Schools have an official curriculum concerning mathematics, but there exist many issues involved in the implementation of the curriculum. The various activities taking place within a classroom and a school are complex and the context and interpretation of the situation have to be taken into consideration. Denzin and Lincoln (2000, p.8) have rightly pointed out that "Qualitative researchers stress the socially constructed nature of reality, the intimate relationship between the researcher and what is studied, and the situational constraints that shape inquiry". Human beings learn by interacting with others and give interpretation to situations based on their experience.

Epistemological assumption

The epistemological belief underlying the study was that there must exist a transactional and subjective relationship between the researcher and the participants. The context in which the study was conducted had to be well understood, and a 'contract' with the participants established. Until that element of trust and mutual respect was established, appropriate data could not be collected. Throughout the research process I was required to make appropriate choices: what to observe, when to be involved with the participants or when to keep my distance; when to create opportunities for interaction or when to refrain from participating. I also had to ascertain that my own beliefs were not influencing my interpretations and explanation of the things noted and observed. This was achieved by establishing a number of quality controls (described later in this thesis).

Methodological assumption

A methodological question deals with how best to pursue the research purpose. How are data to be collected and analysed in relation to the research questions? The purpose of the study was to identify what types of difficulties students in Mauritius encounter while studying mathematics at the secondary level. What are the main factors that affect the learning of mathematics by boys and girls at that level? What is the contribution of teachers, parents, peers and students' own attitude towards mathematics in influencing the learning of mathematics? What strategies can help in enhancing students' mathematics achievement and attitude towards the subject? As mentioned earlier two methods were used, one for the first phase and the other for the last two phases of the study:

Phase one: a survey approach

Phase two: a case study approach.

Phase three: a case study approach.

A survey method was chosen for the first phase because of the need to obtain an overall picture of the situation concerning the teaching and learning of mathematics and related issues across the island. Mathematics has already been described (in Chapter One) to be a subject of great importance to secondary education in Mauritius, and also in the job market. Because of the importance attached to the subject, mathematics is a compulsory subject at the School Certificate level. It was therefore necessary to gain an understanding of the difficulties involved in the teaching of the subject at the national level irrespective of a school's geographical position, type or any other criterion. A stratified method of sampling was used to select the schools to form the sample and the details are provided later in this chapter. Data were collected through the use of two questionnaires, details of which are also provided later. Appropriate statistical analyses were carried out using the Statistical Package for Social Sciences (SPSS).

Once the initial general information was obtained, an in-depth study was carried out. For this phase, a case study approach was used. Four secondary schools were identified to form part of this sample. Ethnographic techniques were used to obtain a holistic view of the problem. These techniques involved observations, interviews and documentary analyses. A case study approach was adopted because, as pointed out by Hitchcock and Hughes (1995, p. 322), the following characteristics of that methodology were found to apply, namely:

- The study called for a rich and vivid description of events relevant to the case
- It blended a description of events with the analysis of them
- It focused on individual persons or groups of people, and sought to understand their perceptions of events
- The researcher was integrally involved in the case (cited in Cohen, Manion and Morrisson, 2000, p. 182).

The appropriateness of this choice is further supported through the following words of Bromley (1986): "…case studies get as close to the subject of interest as they possibly can, partly by means of direct observation in natural settings, partly by their access to subjective factors (thoughts, feelings, desires)…" (cited in Merriam, 1988, p.29).

Following these phases, appropriate strategies to help in enhancing the teaching and learning of mathematics at the secondary level in Mauritius were designed. An evaluation of these recommendations was important and this was the main aim of the third phase. A case study approach was again adopted for this phase. Three secondary schools (one single boys', one single girls' and one coeducational) were selected and a Form IV class in each of them was identified. Mathematics lessons were taught in each of the classes for a period of three months using a teaching and learning package. Pretests and posttests were administered to the students to evaluate the effectiveness of the strategies. Interviews of students were also conducted. Details on the teaching and learning used for this implementation stage are provided in Chapter Six.

The design of the study

The study was developmental, being conducted in three phases. In the first phase, data were collected mainly from two questionnaires: one specially designed for the study and the other one a modified version of the Fennema Sherman Mathematics Attitude Scale. In the second phase data were collected though classroom observations, interviews with students, teachers, parents, rectors and other stakeholders in the educational sector. The two questionnaires used in the first phase also were administered to the different sample of students in the second phase. The students were also asked to fill in the Questionnaire on Teacher Interaction with a view to obtain their perceptions of the type of qualities their teachers possessed and their opinions on the type of teacher interaction present in their class. They were also asked to draw a mathematician and to write down two reasons why they thought that one may need the services of a mathematician. This was carried out in order to find what portrait came to the mind of the students whenever they thought of a mathematician. This Draw-a-mathematician-test is described later in this chapter. In the third phase, strategies developed based on the findings of the two prior phases to enhance the teaching and learning of mathematics at the secondary level were designed through a teaching and learning package and were tested in the three selected schools. A pretest related to concepts of mathematics at the secondary level was administered to the students in the chosen Form IV classes of the selected schools. Mathematics lessons were then taught via the teaching package based on the ASEI (Activity, Student, Experiment and Improvise) movement through the PDSI (Plan, Do, See and Improve) approach. This technique ascertains the full involvement of students in their learning of mathematics. A posttest was then administered to evaluate the efficiency of the strategies. The students were also asked to fill in the What Is Happening In this Class? (WIHIC) questionnaire with a view to obtain their perception on how the mathematics classes were conducted. All the students in the three classes were asked to write a report on their feelings about the happenings in these mathematics classes for the period of three months.

The design of the study is represented in Table 3.1.

Table 3.1: The research design

		Interpretive model of research	
Phase of the study	Method used	Instruments/Techniques used	Sample
One	Survey approach	1. Mathematics Questionnaire	17 schools
		2. Modified Fennema- Sherman Mathematics Attitude Scale	607 students
Two	Case study approach	1. Mathematics Questionnaire	4 schools
		2. Modified Fennema- Sherman Mathematics Attitude Scale	118 students
		3. Classroom Observation	
		4. Questionnaire on Teacher Interaction	
		5. Draw a Mathematician Test	
		6. Interview of students	36 students
		7. Interview of teachers	4 teachers
		8. Interview of parents	12 parents
		10. Interview of other stakeholders	5 in all
Three	Case study approach	1. Pretest	3 schools
		2. Design of the teaching and learning package	111 students
		3. Administration of the teaching and learning package	
		4. WIHIC questionnaire	
		5. Posttest	
		6. Report writing from students	

Quality criteria of the study: trustworthiness, the hermeneutic process and authenticity

Much criticism has been leveled at the case study method regarding the possible lack of rigour and generalisability as compared to quantitative methods. Many criteria have been proposed by researchers to counteract these criticisms. Guba and Lincoln (1989) have pointed out that the quality of interpretive research can be ascertained through trustworthiness, the nature of the hermeneutic process and authenticity.

Trustworthiness is equivalent to the concept of reliability, validity and objectivity in the conventional positivist paradigm (Guba and Lincoln, 1989). It can be ascertained through a number of criteria:

- Credibility: This is parallel to the notion of internal validity in the conventional paradigm and it is an assessment of the "isomorphism between constructed realities of respondents and the reconstructions attributed to them" (Guba & Lincoln, 1989, p.237). In this study the credibility has been observed through prolonged engagement. In fact, in the second phase of the study, regular contacts with the students were made during a period of four months. This helped in establishing a rapport between the students and myself, developed mutual trust and also provided opportunities to confirm or reaffirm certain information. Peer debriefing was also carried out regularly. Frequent discussions were carried out with two colleagues in the same department and one colleague in another department. The questions asked by them at some point in time gave me the opportunity to go back and check certain facts concerning the data collected. Moreover, member checks were carried out to ascertain that the correct messages and information were being received by me and there were no distortions to the facts. All these factors contributed to ensuring the credibility of this study.

- Transferability: This is parallel to the notion of external validity in the conventional positivist paradigm. Transferability is normally established by providing a thick description: "an extensive and careful description of the time, the place, the context, the culture in which those hypotheses were found salient" (Guba & Lincoln, 1989, pp.241-242). The main idea is to provide a rich and as complete database as possible with a view to helping a reader reach a decision concerning the transfer of these findings to his/her situation. In this present study, data were collected over a period of time and through various methods. The amount of information collected ensured that a thick description of the situation could be made.

- Dependability: This is parallel to the notion of reliability in the conventional paradigm – that is, the extent to which a test or procedure produces similar results under the same conditions. Thus, dependability is concerned with the stability of data over time.

- Confirmability: This is parallel to the notion of objectivity in the conventional paradigm. "Like objectivity, confirmabiltity is concerned with assuring that data, interpretations, and outcomes of inquiries are rooted in contexts and persons apart from the evaluator and are not simply figments of the evaluator's imagination" (Guba & Lincoln, 1989, pp.242-243). I did ascertain that data were collected to reflect the genuine feelings of the respondents. While conducting classroom observations it was assured that my presence in the class would have minimal effect on the classroom discussions. The students were conscious of my presence initially, but after a while my presence attracted no attention at all. This meant that the students behaved in their natural settings and a 'true' picture could be observed. Notes were taken in a journal, and a resume of the happenings in the class was made the same day. Interviews were audio taped and the students were asked to answer in any language they preferred. This was to ascertain that the students felt free to express themselves, and that language would not be a barrier.

The Hermeneutic Process: another criterion that ascertains quality in interpretive models of research is the process itself. Data collected could be analysed at the time of reception and then fed back to the respondents for comment, elaboration, justification, rectification or any other such thing. Possible misconceptions could be sorted out and this further enhanced the quality of the research.

The Authenticity Criteria: Dealing with issue of authenticity, there should be fairness (that is no bias) in the study. To ascertain that, I ensured that a complete and balanced representation of the multiple realities of a situation, as stipulated by Cohen, Manion and Morrison (2000) was achieved. The different responses to each and every issue raised during the study were given due consideration and used to provide a broad picture of the situation.

Phase One: Gaining an overall picture

As already discussed, the study was conducted in three phases. Consequently there were three samples involved. In the first stage, a sample of secondary schools across Mauritius was selected. It should be pointed out that in Mauritius there are three types of secondary schools:

- State: managed by the Ministry of Education and Scientific Research
- Confessional: administered by religious bodies
- Private: privately owned but under the control of the Private Secondary School Authority.

It should also be mentioned that a number of fee-paying schools at the secondary level exist in Mauritius, but as they are relatively new and because they offer a different curriculum from those offered by almost all other schools in Mauritius, this types of school is not included in this study.

The number of schools of each category as they existed in 2003 is shown in Table 3.2.

Table 3.2: Number of schools of each category in Mauritius

Type of school	State	Confessional	Other private
	46	29	65

Taking the specificity of the Mauritian education system into consideration, the following criteria also were established while selecting the schools in the sample:

• Location of school (rural or urban).

• Gender of the students of the school (single male, single female or co-ed.)

Following the Educational Reform in 2002, Mauritius has been divided into 4 zones. Prior to 2002, admissions to a secondary school were made nationally based on the results of the Certificate of Primary Education examinations. This was found to apply considerable stress on the students (10-11 years old) and their parents. So a policy decision was taken by the Ministry of Education and Scientific Research in Mauritius to abolish ranking at the CPE level and to replace it with a grading. Another major change in the reform was that admissions for Form One level were to be made at a zonal level. In Figure 3.1 the different zones with the state secondary schools are shown.

MAURITIUS

State Secondary Schools (Form I - V) 2003

① Port Louis and North
② Beau Bassin - Rose Hill, Centre and East
③ Curepipe and South
④ Quatre- Bornes, Vacoas - Phoenix and West
⑤ Rodrigues

Secondary Schools (Form I - V) ▲ Existing
■ New
B Boys
G Girls

RODRIGUES

Figure 3.1: The Secondary Schools in the Republic of Mauritius following the Reform

The number and type of schools in each zone selected for the sample is shown in Table 3.3.

Table 3.3: Type of schools in each region for the final sample

Region	State	Confessional	Other private
Zone 1	1	0	1
Zone 2	2	2	1
Zone 3	2	0	3
Zone 4	1	2	1
Rodrigues		1	

The number of students involved in this first phase of the study was 606, that is 288 boys and 318 girls.

The instruments used

Two instruments were used to collect data in the first phase of the study. One questionnaire aimed to gain information about mathematics based on the lower secondary curriculum and another to assess the attitude of students towards mathematics.

Questionnaire on mathematics (see Appendix Two)

In order to assess and compare the students' achievement in mathematics in the different schools, a questionnaire was specially designed for the study. In the first instance, the syllabi of mathematics for Form I, II and III were analysed and the different objectives that need to be met by students of these forms were identified. The main aim in designing the questionnaire was to find out how much, and to what extent, the concepts of mathematics of the lower secondary level the students of Form IV had acquired. The strands chosen were Number and Operations, Algebra, Geometry and Data Analysis and Probability. These are four of the five strands advocated by the NCTM Standards (1989). It should be pointed out that more emphasis was given to Algebra in this questionnaire as the students in Mauritius are introduced to Algebra only at lower secondary level and it can be said to be the backbone of mathematics learning. In fact, the place and importance of Algebra in the secondary school curriculum has been emphasized by NCTM (1989). Items based on these strands were written, but in a somewhat different way to the manner students were familiar with in their regular examinations. Items were designed to test the conceptual understanding of the students.

The questionnaire consisted of three parts: the first part dealt with background information concerning the students. It included questions concerning the prior performance of the student in mathematics; the average number of hours devoted to mathematics in a week; information about the parents' education and job; languages the student could speak; the language most often spoken by the student, and the persons having the most influence on the students work in mathematics. The second part consisted of ten multiple choice questions based on the curriculum of lower secondary mathematics, while the third part consists of 13 word problems on the different curriculum strands. The first ten were short answer questions while the last three were more structured. The students were expected to display their conceptual understanding of different parts of mathematics by presenting well-structured solutions. The answers were assessed, but at the same time, the way the student tackled the problem was also noted. Any misconceptions involved were noted to provide an in-depth analysis of students' problem-solving abilities. Before finalizing the questionnaire it was piloted with a batch of 60 students from Form IV randomly chosen. Following the feedback, minor changes were made to the questionnaire for the final version.

I visited all schools personally to distribute the questionnaire forms. The main aim of the survey was clearly explained to the Head of the Mathematics Department of each school together with the Head of the institution. Forty questionnaires were then handed over to the Head of Department and the way to conduct the test was discussed. It was decided that the selected class should be representative of the students in Form IV in the school. It also was stressed that the test was to be conducted in examination conditions to ensure students approached the questionnaires in a serious way. The heads of department of co-educational schools in the survey also were asked to ensure that there were, as far as possible, an equal number of boys and girls in their samples.

Once the completed questionnaires concerning the items on mathematics were obtained, they were checked by me, and the ways the students dealt with the different questions were noted. The responses of the students were recorded using the Statistical Package for Social Sciences (SPSS) and different statistical tests (described later) were performed.

It should be mentioned that for this present mathematical test, 'face validity' was ensured. In fact, the designed questionnaire was discussed with three colleagues in my department who have a rich experience in teaching at secondary schools and also at teacher training institution. The questionnaire was also discussed with an expert in the field of mathematics education. As for reliability of the instrument, it has been used at two different times (and at the pilot stage also) and the schools of comparable standards were almost the same. Cronbach reliability coefficients of 0.62 and 0.50 were obtained for the multiple choice section of the questionnaire for Phase One and Two respectively.

The attitude questionnaire (see Appendix Three)

Attitude towards mathematics plays an important role in mathematics achievement of students. Maker (1982, cited in Ma & Kishore, 1997, p.26) highlighted the relationship between the affective and the cognitive domains: "It is impossible to separate the cognitive from the affective domains in any activity... the most important is that there is a cognitive component to every affective objective and an affective component to every cognitive objective". The importance of incorporating affective factors with cognitive factors has been very much emphasized (NCTM, 1989). In another study Gomez-Chacon (2000, p. 166) pointed out

> ... in investigations on mathematics and the affective dimension, we should address the subject's two affect structures, the local and the global. This last implies viewing the person in his situation, knowing the individual's system of beliefs (beliefs as a learner of mathematics, beliefs about mathematics, beliefs about the school context), social representations and the process of construction of the subject's social identity.

Many studies have been conducted to analyse the attitude of boys and girls towards mathematics, and a seminal questionnaire – the Fennema-Sherman Mathematics Attitude Scale – was designed in 1976. It consists of nine scales namely: Confidence in Learning Mathematics; Effectance Motivation; Mathematics Anxiety; Usefulness of Mathematics; Attitude toward Success in Mathematics; Mathematics as a Male Domain; Father; Mother; and Teacher. This questionnaire has been used extensively (Fennema, 2000) in many studies related to the attitude of students towards mathematics. However in some recent research (Forgasz, Leder, & Gardner, 1999) it has been reported that some of the items, especially in the scale Mathematics as a Male Domain, may not be quite valid nowadays and may require amendment.

For this study the *Modified Fennema-Sherman Mathematics Attitude Scale* questionnaire was used. It consists of 47 items dealing with four subscales:

Confidence, Usefulness, Mathematics as a Male Domain and Teacher Perception. In each subscale items are worded with a positive or a negative connotation. Students are expected to provide their responses to each of these items using a five point response format ranging from Strongly Agree, Agree, No Opinion, Disagree to Strongly Disagree. For the purpose of the analysis positively framed items were marked as follows:

St. Agree- 5; Agree- 4; No opinion- 3; Disagree- 2 and St. Disagree- 1,

and negatively worded items in the reverse way; that is:

St. Agree- 1; Agree- 2; No opinion- 3; Disagree- 4 and St. Disagree- 5.

The sum of the marks for the items under each scale were computed and thus for each student there were four scores (one for each scale) and a total score. The mean and standard deviation were computed and used for further analysis.

Phase Two

The purpose of the second phase of the study was to delve deeper into the issues noted during the first phase. For the second part, four schools were selected: one single boys', one single girls' and two co-educational. A letter, in which the main objectives of the study were explained and the permission sought to conduct this phase of the research, was written to each rector in October 2003. Once the permission was granted, I visited each of the selected schools, and met the rector and discussed further details concerning the study. A Form IV class from each school was identified with the help of the rector and the Head of the Department of mathematics while the agreement of the classroom teacher to participate was also sought. During the first visit to each class, the purpose of the study, together with the objectives, were made clear to the students. They were informed about the confidentiality of all the information that they would provide. Once the agreement of the children was obtained, classroom observations were carried out to analyse the different interactions taking place within the mathematics classroom.

Obtaining data through observation

Classroom observations were conducted in each of the classes. A nonparticipation observation approach to the study was used where the students were observed in their natural settings (Fraenkel & Wallen, 1993). Students "live" together and consequently act and interact between themselves and with others. It is of paramount importance to analyse this phenomenon more closely as "human beings act towards things on the basis of the meanings they have for them" (Woods, 1979, cited in Cohen, Manion & Morrison, 2000, p. 25). To determine what problems boys and girls are facing concerning mathematics, one should try to find out what interpretations of the different transactions in the classroom boys and girls have and how these may affect their attitude and consequently performance in mathematics. As has been pointed out by Cohen et al. (2000): "interactionists focus on the world of subjective meanings and the symbols by which they are produced and represented" (p. 25). The importance of observation has been highlighted by Denscombe (1998): "Observation offers the researcher a distinct way of collecting data. It does not rely on what people say they do, or what people say they think. It draws on the direct evidence of the eye to witness events first hand" (p. 139).

Different things happening within mathematics classes were observed: for example, teacher interactions with boys and girls in the coed class; the number and type of questions asked by a boy/girl in a class discussion; the peer discussion that goes on in a class; the interest shown by students in mathematics and mathematical activities and other such related issues. An observation schedule was specially designed for this purpose (see Appendix Four), and observations were recorded in a journal on the basis of this schedule. Any other event happening or behaviour that was relevant to the study was also noted. A detailed summary description of what was observed in the classroom was normally made by me in the evening from the notes jotted down during the day. (See Appendix Thirteen for example of descriptions). In certain cases, audiotaping was used to ensure that I did not miss any important information.

The classroom observations were then followed by the administration of the mathematics and attitude questionnaires to the students of each of the four classes. The same statistical tests conducted with the first sample of students were then applied.

Generating data through interviews

Following the observation sessions, it was imperative to conduct semi-structured interviews with selected students to probe deeper and find out more information.

"Interviewing is an important way to check the accuracy of – to verify or refute – the impressions he or she has gained through observation" (Fraenkel & Wallen, 1993, p. 385). It should be noted that qualitative researchers are concerned with process as well as product. The teacher in each of the four schools were asked to identify nine students in all based on the following criteria:

3 high achieving students in mathematics

3 average students in mathematics

3 students facing difficulties in the learning of mathematics.

The teachers were asked to use the mathematics results from the first term and their own knowledge of the students to make the choice. An equal number of boys and girls were selected for the final sample for interviews. Once selected, all the nine students in a school were interviewed as a group. Each one was asked to respond in turn to each of the questions (see Appendix Five for the interview sheet). The students were reassured of the confidentiality of their opinions and responses to the issues raised. Notes were made about what students said which was supported by audio recording (with their permission). This helped me while interpreting the interview on issues that I may have missed in listening to responses. During the interviews I ensured that I asked the question and prompted in certain cases, but always remained non-judgmental. I made a summary of the main points raised during the interview to validate the responses of the students. An example of an interview transcript and notes is provided in Appendix Six.

Teachers, rectors, parents and other stakeholders were given the opportunity to express their ideas on the teaching and learning of mathematics and gender issues. Interviews were conducted with the four teachers who formed part of the sample for the second phase of the study. Their responses were audio taped and notes also made in a journal. Interviews also were conducted with the rectors of the four schools. Their vast experience in secondary schooling proved to be of utmost importance for this study. They had worked with different cohorts of students over a long period of time and they contributed considerably in explaining the attitude of students towards mathematics and schooling at large, as well as other related issues. Parents too were interviewed to determine their views concerning the involvement of their children in mathematics learning and other activities related to schools. In fact for each school, one parent corresponding to each of the three groups of students interviewed (high achieving, average and those having difficulties in mathematics) was chosen, and interviews were conducted in their homes. In all, twelve parents were involved in the interview process.

It was also important to have the views of other stakeholders on the issue of gender and the learning of mathematics, consequently three retired mathematics teachers with long experience in secondary schooling and two lecturers in a tertiary institution were interviewed. All the interviews were audio taped.

Generating data on teacher interaction

Many questionnaires have been designed for assessing students' perception of the classroom environment (Fraser, 1998). To have an idea of the type of teacher interaction present in each class participating in this study, the Questionnaire on Teacher Interaction (QTI) was administered to all the students in the four classes. Teacher interaction plays a very important role in establishing a positive environment in a class. Research has been carried out on the nature and quality of interpersonal relationships between teachers and students (Wubbels & Brekelmans, 1998; Wubbels & Levy, 1993). The importance of assessing the learning environment in the teaching and learning process has also been emphasized in many research studies (Fraser, 1998). It has been noted that students' perceptions of their classroom psychosocial environment have a significant influence on their cognitive and affective learning outcomes (McRobbie & Fraser, 1993; Yarrow, Millwater, & Fraser, 1997).

Based on a theoretical model of proximity (cooperation-opposition) and influence (dominance-submission), the QTI was developed to assess student perceptions of their interactions with their teacher. These dimensions can be represented in a coordinate system divided into eight equal sectors (Wubbels, 1993).

The sectors are labeled DC, CD, etc depending on their position in the coordinate system. For instance, the two sectors CS and SC are both characterized by Cooperation and Submission. In the CS sector, the cooperation aspect prevails over the submission aspect, while in the SC sector it is vice versa.

The eight scales involved in the QTI are:

- Leadership,

- Understanding,

- Uncertain,

- Admonishing,

- Helping/Friendly,

- Student Responsibility and Freedom,

- Dissatisfied and

- Strict.

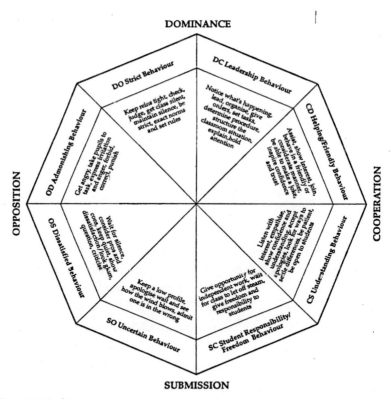

Figure 3.2: The Model for Interpersonal Behaviour (Wubbels, 1993)

The questionnaire consists of 48 items, six on each of the scales. Students are expected to respond to each of the item concerning the interaction with his/her mathematics teacher. The total score for each of the scale is computed for each student and the mean, standard deviation and Cronbach reliability coefficient are found for each of the scales.

The data obtained helped in analyzing the type of teacher interaction present in the four classes chosen for the study. After a first analysis of the responses, mean, standard deviation and alpha Cronbach reliability coefficient were computed. A sample of the students in each class was then interviewed to probe deeper into each of the QTI scales.

The Draw-A-Mathematician-Test

The perception that one has of mathematics and a mathematician may influence one's participation and performance in the subject, as noted below:

> [I]t is a matter of great concern that ... negative images of mathematics might be one of the factors that has lead to the decrease in student enrolment in mathematics and science at institutions of higher education, in the past decade or two... the term 'image of mathematics' refers to a mental picture, view or attitude towards mathematics, presumably developed as a result of social experiences, through school,, parents, peers, mass media or other influences. (cited in Picker & Berry, 2000, p. 2)

The use and efficiency of images and drawings to provide glimpses of students' perceptions about mathematicians and mathematics has been commented on by many researchers such as Picker and Berry (2000; 2001; 2002). These researchers have suggested that imagery can provide insight into the belief systems, assumptions, and expectations that students hold in relation to mathematics. A well-known instrument called the Draw-A-Scientist-Test (DAST) (Chambers, 1983) has been used extensively in many parts of the world to reveal the perceptions students held about scientists and science. This test also has been used, with modifications, to examine students' ways of thinking about mathematicians and mathematics.

The use of drawings to elicit students' ideas and developmental achievements has a long history (Gardner, 1980, 1993; Goodenough, 1926; Goodenow, 1977; Harris, 1963; Pedersen & Thomas, 1999). Drawings have been used extensively by psychologists for a range of purposes such as assessing children's motor and cognitive development (Gardner, 1980), examining the content of children's drawings in relation to cultural context (Krampen, 1991) and to help children deal with social and emotional trauma such as abuse (Wakefield & Underwager, 1998).The anthropologist Margaret Mead and psychologist Rhoda Métraux conducted a study using a composite of features attributable to scientists to investigate the images of scientists held by American high school students (Mead & Metraux, 1957). Based on this work, and using the seven most significant features students in the study nominated, Chambers (1983) devised the Draw-A-Scientist Test (DAST). The DAST was used to investigate young children's perceptions of scientist because it was believed that their ideas could not be expressed fluently through writing.

Inspired by the work of Chambers and others concerning DAST, researchers in mathematics education have tried to find out the perception of students concerning a mathematician. Furringhetti (1993) rightly pointed out that mathematics "is a discipline that enjoys a peculiar property: it may be loved or hated, understood or misunderstood, but everybody has some mental image of it" (cited in Picker & Berry, 2000, p. 65). It is extremely important to understand the perceptions that students hold concerning mathematicians and mathematics to enable the teaching and learning of mathematics to be carried out effectively. In fact, Lim and Ernest (1999, cited by Picker & Berry, 2000) emphasised that teachers should ascertain how popular or unpopular mathematics is with their students to be able to find ways to change and improve the image of the subject and students' attitudes towards it.

To gain an idea of the perception of the students in the four schools chosen for the study concerning a mathematician and mathematics, the Draw-A-Mathematician-Test was administered to them. Each student in the classes chosen for the second phase of the study was asked to draw a mathematician at work. They were also asked to give three reasons why one would need the services of a mathematician. Through this activity the intrinsic attitude and perception of the student towards mathematics could be determined which in some way or the other could have been influencing their response and engagement towards mathematical activities.

Generating data through the examination of materials

Further data were collected through the examination of different materials. The copybooks of students did provide an opportunity to browse the 'thinkings' and 'ways of working' of the students as it was possible to analyse what processes boys and girls were using while solving a problem. The scripts of students for internal examinations were also analysed. All these documents helped in gaining insights concerning the students' attitude and motivation towards mathematics. Analyses of Cambridge Examination Reports were also conducted to identify the type of mistakes and difficulties students encountered at the School Certificate Examinations. These analyses provided issues to be raised during interviews.

Bringing together the different data

The different data collected through the classroom observations, questionnaires administered, semi-structured interviews of students, teachers, rectors, parents, and other stakeholders and from document analysis were analysed with respect to the different research questions. Patterns and similarities in the issues raised were explored. Analyses were carried out as suggested by Miles & Huberman (1994) namely: data reduction, data display and conclusion drawing/verification. Data reduction "refers to the process of selecting, focusing, simplifying, abstracting, and transforming the data that appear in written-up field notes or transcriptions" (Miles & Huberman, 1994, p.10). Selected data were then displayed in a way to allow drawing of conclusions and actions. The issues emerging from the data were then tested "for their plausibility, their sturdiness, their 'confirmability' – that is their validity" (Miles & Huberman, 1994, p. 11). It should however be pointed that examining three qualities did in fact form an interactive, cyclical process.

Phase Three

Phase Three was the implementation phase. Strategies that were identified after having collected data from phases one and two of the study were implemented in three secondary schools (one single boys', one single girls' and one coeducational) for a period of three months. The three schools that were selected were of comparable standards in terms of performance at the School Certificate level and in terms of reputation in the public. Initial meetings were conducted with the rector of each school where the objectives of the study and this implementation phase was clearly explained. After having obtained the permission of the rector, a meeting with the Head of mathematics department was held and again the objectives of the study and the implementation programme were explained. The head of department then chose one Form IV class and a meeting with the classroom teacher was held to explain again what was planned to happen in his class and seek his permission. The scheme of work was discussed with the teacher and the chapters that I would deal in my sessions with the students were identified. The selected chapters were chosen from the scheme of work of the first term − namely: Quadratic Equations, Applications of Quadratic Equations and Angle Properties of Circles. In each school I taught the identified class once per week for 40 minutes over a period of three months. A detailed description of the pre-test and post-test used for this phase of the study, together with the strategies that were used in the classrooms, is given in Chapter Six. To investigate the perceptions of the students involved in this phase of the study about the different issues like teacher support, cooperation and equity, the WIHIC questionnaire was administered to them Details about this questionnaire and other strategies that were used within the classes in this phase of the study are discussed in Chapter Six.

Ethical Issues

The inclusion and consideration of ethical issues now has a greater emphasis in educational research than it did ten or fifteen years back. "The awareness, focusing chiefly, but by no means exclusively, on the subject matter and methods of research in so far as they affect the participants, is reflected in the growth of relevant literature and in the appearance of regulatory codes of research practice formulated by various agencies and professional bodies" (Cohen et al., 2000, p. 49). Proper consideration was given to the ethical obligations during the conduct of the study. The following issues were addressed:

All the students, teachers, rectors, parents and other stakeholders were informed of the above issues concerning this study. Permission was sought before involving anyone in the research. Letters were written to the Permanent Secretary of the Ministry of Education and Scientific Research explaining the objectives of the research and asking permission to conduct the study in the state schools. Once permission was granted, direct contact with the rector of each school was established where again the objectives of the study were discussed. For the other schools, letters were written directly to the rector. The class where the study could be conducted was identified with the help of the rector and the permission of the class teacher was sought to conduct the study. He/she in turn discussed with the students and explained the objectives of the study and sought their permission to be involved. The students were also informed that they might withdraw from the study at any time without prejudice. I assured the participants (individuals or organizations) that strict privacy and confidentiality of the information collected would be maintained − a strategy that I believe further motivated the participants to speak their mind and give genuine and trustworthy information about the various issues. I ensured that disruption of the normal routine of a school or a class was kept to a minimum, and that my presence in the classroom did not have a negative impact in the usual classroom daily routine.

Triangulation

Triangulation is an important component of educational research and it is used "to enhance the validity of research findings" (Mathison, 1988, p. 13). As already discussed, many steps were taken during the course of this study to ensure validity. Data were collected from different sources and at different times and places which brought about data triangulation. Also different methods were used to collect data and this helped in bringing about methodological triangulation. Investigator triangulation also was catered for to some extent as discussions were carried out with colleagues at regular intervals. The following words of Mathison (1988) explain well the role and importance of triangulation. "The value of triangulation lies in providing evidence − whether convergent, inconsistent, or contradictory − such that the researcher can construct good explanations of the social phenomena from which they arise" (Mathison, 1988, p. 15}. All the data collected in both the phases of the study have assisted in obtaining a holistic view of the problem and establishing a systematic way of working the solution. "Theorizing is the act of constructing from data an explanatory scheme that systematically integrates various concepts through statements of relationship" (Strauss & Corbin, (1998, p.25).

Summary of the chapter

In this chapter, the methodology that was adopted for this study has been discussed and the interpretive model of research chosen for this study described. The different phases that constituted the study were discussed in detail together with the quality criteria. The different samples for the three phases of the study were also described together with the various instruments that were used. The chapter ended with an explanation of the ethical considerations that were taken throughout this study.

The analyses of data collected in the first phase of the study are discussed in the next chapter.

CHAPTER FOUR
Analysis 1

In this chapter the responses of the students to the different questions in the mathematics test are analysed. The test was divided into three parts: the first dealt with background details of the student; the second (Section A) consisted of a series of 10 multiple-choice questions and the third (Section B) consisted of word problems in mathematics. It should be noted that multiple-choice questions are not set in examinations at the secondary level in Mauritius. The questions in Section A and Section B were set in such a way to test the conceptual understanding of students concerning mathematics at that level. (A copy of the questionnaire is in Appendix Two). The analysis was carried out in three phases using descriptive statistics, comparative statistics and regression analysis respectively.

Section A
In this section each question is presented, followed by a table indicating the different possibilities and the percentage response for each one. The correct answer is marked with an asterisk.

Question 1: **The place value of 6 in 56824 is:**

Table 4.1: Response to Question 1

Possibilities		% response
A	6	0
B	60	0.7
C	600	0.8
D *	6000	98.2
No response		0.3

This question was correctly answered by almost all the students in the sample – 98.2 %. Place value is a concept which is taught to students from their early primary days. Undoubtedly students have mastered this concept.

Question 2: Which of the following figure has rotational symmetry but does not have line symmetry?

Table 4.2: Response to Question 2

Possibilities		% response
A	A square	4.1
B*	A parallelogram	47.4
C	A rhombus	35.7
D	An equilateral triangle	7.6
No response		5.2

Note that around 50% of the students were unable to answer this question correctly. Line symmetry has been taught since primary grade, while rotational symmetry was a relatively new concept for the students as it was introduced at the end of the lower secondary level (Grade 8). It has been found that the concept of rotational symmetry is a matter of concern for many students at the secondary level. Most of the students who answered this question incorrectly identified the **rhombus** as the figure which has rotational symmetry but does not have line symmetry instead of the **parallelogram**.

It was also noted that there was little difference in the responses of boys and girls in this question. Figure 4.1 shows the response of boys and girls graphically.

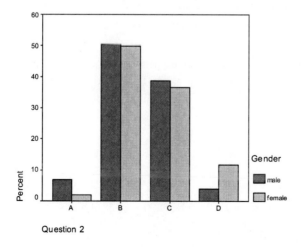

Question 2

Figure 4.1: Response to Question 2 gender wise

A Mann-Whitney U test concerning the responses of boys and girls for this question produced a p-value of 0.824 which showed that there was no significant gender difference in the responses of students.

Question 3: **The probability that a boy is late for school in a given day is $\frac{1}{5}$. The probability that the boy is late for school in two consecutive days is:**

Table 4.3: Response to Question 3

Possibilities		% response
A	equal to $\frac{2}{5}$	57.2
B	equal to 0.2	2.3
C*	less than $\frac{1}{5}$	24.8
D	more than $\frac{2}{5}$	12.0
No response		3.7

Note that approximately two third of the sample of students could not answer this question correctly. The most common answer to be found was "$\frac{2}{5}$", indicating a misconception concerning probability: these students have **added** the two probabilities instead of **multiplying** them.

Considering this question from a gender perspective, there was a statistical difference between the responses of boys and girls.

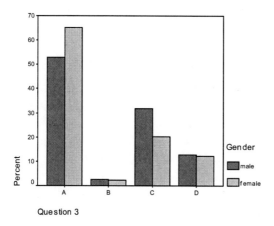

Question 3

Figure 4.2: Response to Question3 gender wise

A Mann-Whitney U test yielded a p-value of 0.001 which showed that a significant difference exists in the responses of students regarding gender.

Question 4: **Which of the following statements about zero is false?**

Table 4.4: Response to Question 4

Possibilities		% response
A	$2 \times 0 = 0$	1.0
B	$\dfrac{0}{2} = 0$	12.8
C*	$\dfrac{2}{0} = 0$	82.4
D	$2 + 0 = 2$	2.5
No response		1.3

The majority of the students answered this question correctly. However, 12.8 % of the students identified the statement "0/2 = 0" as false instead of the statement "2/0 = 0".

A graphical representation of the responses of boys and girls to this question is given in Figure 4.3.

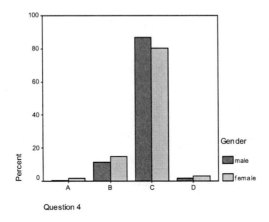

Figure 4.3: Response to Question 4 gender wise

A Mann-Whitney U test yielded a p-value of 0.026 which showed that there is a significant gender difference in the responses of students.

Question 5: **The number of significant figures in 52.003 is**

Table 4.5: Response to Question 5

Possibilities		% response
A	1	0.7
B	2	11.3
C	3	29.9
D *	5	55.8
No response		2.3

Almost half of the students were unable to answer this question correctly. Approximately 30 % of the students identified the number of significant figures as "3" instead of "5". It was also found that the responses of boys and girls for this question were not significantly different (p= 0.026 for the Mann- Whitney U test).

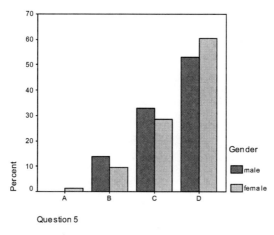

Question 5

Figure 4.4: Response to Question 5 gender wise

Question 6: **If X and Y are two non-empty sets (X, Y ⊂ ξ) such that X ∩ Y = Y, then**

Table 4.6: Response to Question 6

Possibilities		% response
A	$X = Y$	9.4
B*	$Y \subset X$	38.2
C	$X \subset Y$	19.2
D	$X \cup Y = \xi$	20.2
No response		13.0

Approximately 38 % of the students were able to provide the correct answer to this item. Sets form part of the curriculum of mathematics from primary level and students are very much at ease with this chapter of the textbook as far as the usual questions set in the examinations are concerned. Note that 20.2 % of the students identified the answer "$X \cup Y = \xi$" showing that they did not pay attention to the fact that it was specified in the question that the two sets X and Y are proper subsets of the universal set.

An analysis revealed that there was no significant difference between the responses of boys and girls for this question. (p-value = 0.158 for the Mann-Whitney U test).

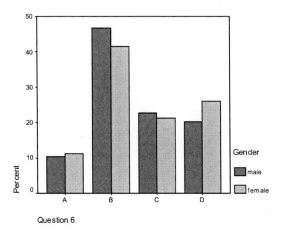

Figure 4.5: Response to Question 6 gender wise

Question 7: **If p − 145 = 225, then p − 144 =**

Table 4.7: Response to Question 7

Possibilities		% response
A	223	0.5
B	224	16.9
C	225	1.0
D *	226	79.9
No response		1.7

About 80% of the students responded correctly to this question. When marking the scripts it was noted that the vast majority first calculated the value of p and then subtracted 144 from this value. Only a few could argue that if p − 145 = 225, then p − 144 should be 1 unit greater than 225 and thus 226.

There was a significant difference between the response of boys and girls to this question. as confirmed by a p-value of 0.002 for the Mann-Whitney U test.

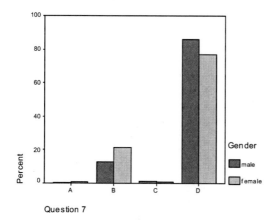

Question 7

Figure 4.6: Response to Question 7 gender wise

Question 8: $1 - \dfrac{1}{3} \times \dfrac{1}{4} =$

Table 4.8: Response to Question 8

Possibilities		% response
A	$\dfrac{1}{6}$	23.0
B	$\dfrac{1}{2}$	2.6
C	$\dfrac{10}{12}$	2.1
D *	$\dfrac{11}{12}$	70.9
No response		1.4

Approximately 30% of the students could not answer this question correctly. This can be considered quite high as this concept is basic and has been taught from Form I. The incorrect answer which was found to be given by most of the students was "$\dfrac{1}{6}$", showing that these students first computed $1 - \dfrac{1}{3}$ to get $\dfrac{2}{3}$ and then multiplied by $\dfrac{1}{4}$.

An analysis of the responses of boys and girls showed that there was no significant difference in their answers (p-value = 0.269 for the Mann-Whitney U test).

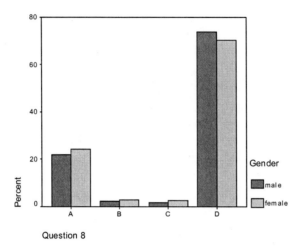

Figure 4.7: Response to Question 8 gender wise

Question 9: **Samantha (S) and Tina (T) walk towards each other. Samantha walks on a bearing of 140°. The bearing on which Tina walks is**

Table 4.9: Response to Question 9

Possibilities		% response
A	40°	52.5
B	220°	13.2
C	310°	0.7
D*	320°	30.3
No response		3.3

Approximately 30% of the students could provide the correct answer. It has been highlighted in different discussions and several examination reports that bearing is a topic in which students tend to face difficulties. It should be mentioned that 52.5 % of the students gave the answer "40" instead of "320" which shows that these students have problems with the basic concept of bearing and that the angle should be measured in a clockwise direction.

The responses of boys and girls for this question were significantly different. A p-value of 0.000 was obtained for the Mann-Whitney U test.

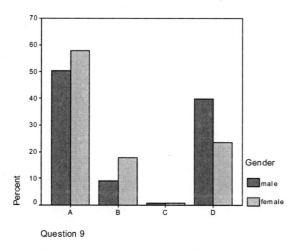

Question 9

Figure 4.8: Response to Question 9 gender wise

Question 10: **Which of the following statements is false about the mean of a set of numbers?**

Table 4.10: Response to Question 10

Possibilities		% response
A*	It is always a positive number.	41.9
B	It cannot be found if one of the numbers is missing.	17.4
C	It uses all the data in the set.	7.7
D	It is affected by extremely large or extremely small values in the set.	18.3
No response		14.7

Approximately 42% of the students answered this question correctly. Finding the average of a set of scores is in the upper primary level mathematics school curriculum and students normally do these problems correctly. The response to this question shows that approximately two-thirds of the students lack a proper understanding of the concept. The two most common statements which were identified to be wrong were **"It cannot be found if one of the numbers is missing"** and **"It is affected by extremely large or extremely small values in the set".**

There was no significant difference between the responses of boys and girls for this question (a p-value of 0.706 for the Mann-Whitney U test).

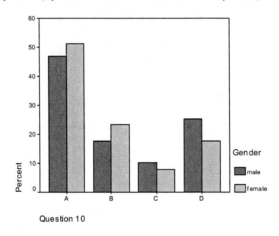

Question 10

Figure 4.9: Response to Question 10 gender wise

Section B

Question 11(i): **Solve the equation** $\dfrac{1}{2}x + 3 = \dfrac{1}{3}x - 5$

Objective: To see how students solve a linear equation involving fractions.

Table 4.11: Response to Question 11(i)

Full marks	Mean	Standard deviation
3	2.12	1.23

61.3% of the students obtained full marks. However 20.9% scored zero marks in this question. This is a high percentage and needs attention. Most of the students manipulated the equation as it is with the fractions while very few eliminated the fractions by multiplying by the LCM of 2 and 3.

An analysis gender-wise revealed no significant difference between the responses of boys and girls. A Mann-Whitney U test gave a p-value of 0.234 which showed that there is no significant gender difference in the responses of students.

Table 4.12: Mean for boys and girls in Question 11(i)

Mean for boys	Mean for girls
2.17	2.08

Question 11(ii): **Solve the equation $x^2 + x = 6$.**

Objective: To identify the different strategies for solving a quadratic equation.

Table 4.13: Response to Question 11(ii)

Full marks	Mean	Standard deviation
3	2.08	1.32

Almost 66% of the students scored full marks. This concept is relatively new as it is introduced for the first time in Form 3. However it is a basic one for the mathematics as well as the Additional Mathematics Curriculum. Also, 26% of the students scored 0 marks for this question.

An analysis revealed no significant difference between the responses of boys and girls (a p-value of 0.885 for the Mann-Whitney U test).

Table 4.14: Mean for boys and girls in Question 11(ii)

Mean for boys	Mean for girls
2.07	2.08

Question.12: **Without using calculator evaluate $6.27^2 - 3.73^2$**

Objective: To test whether students can apply the concept of difference of two squares in arithmetic problems.

Table 4.15: Response to Question 12

Full marks	Mean	Standard deviation
3	1.40	1.40

40.0% of students scored full marks for this question while 46.6% scored zero marks. It should also be noted that many students have carried out the actual multiplication and then the subtraction. Another interesting finding in the responses of the students is the answer " 2.54^2 " showing the misconception " $a^2 - b^2 = (a - b)^2$ " .

An analysis gender wise showed that the responses of boys and girls to this question were not significantly different (a p-value of 0.331 for the Mann-Whitney U test).

Table 4.16: Mean for boys and girls in Question 12

Mean for boys	Mean for girls
1.45	1.34

Question 13: **A quantity is increased by 20% but later decreased by 10%. What is the overall percentage change?**

Objective: To test the conceptual understanding of students concerning percentages.

Table 4.17: Response to Question 13

Full marks	Mean	Standard deviation
4	0.29	0.95

This question can be considered to be amongst the most difficult ones in this mathematics questionnaire as only 4.6% of the students could score the total marks and 89.6% scored zero. Almost all the students who were incorrect gave the answer "10%", showing that they did not really understand the concept of percentage. They calculated "20% – 10%= 10%". Note that students are working with the concept of percentage since upper primary level.

Considering this question from a gender perspective there was a statistical difference between the responses of boys and girls. A Mann-Whitney U test yielded a p-value of 0.000 was obtained which showed that there is a significant difference in the responses of students regarding gender.

Table 4.18: Mean for boys and girls in Question 13

Mean for boys	Mean for girls
0.48	0.13

Question 14: **Given that** $\sqrt{50}$ = 7.071, $\sqrt{5}$ **= 2.236, without using calculator, find the value of** $\sqrt{5000}$.

Objective: To test whether students can apply their understanding of square roots.

Table 4.19: Response to Question 14

Full marks	Mean	Standard deviation
4	2.68	1.83

Note that 64.1% of the students scored full marks while 29.3% scored zero.

An analysis of the mean performance of boys and girls in this question revealed that there was not much difference between them. However, a Mann-Whitney U test yielded a p-value of 0.014 which showed that there is a significant gender difference in the responses of the students.

Table 4.20: Mean for boys and girls in Question 14

Mean for boys	Mean for girls
2.86	2.51

Question 15: **ABC is a right-angled triangle with AB = 8 cm; BC = 6 cm and ABC = 90°.**

Find

(i)   **AC**

(ii)   **Area of** $\triangle$**ABC**

(iii) **BX**

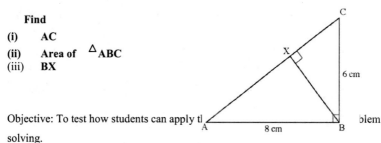

Objective: To test how students can apply th... blem solving.

Table 4.21: Response to Question 15

Full marks	Mean	Standard deviation
4	2.02	1.17

Around 19% of the students scored full marks in this question while 53.4% scored 2 marks. Those who scored 2 marks could complete the first two parts correctly which involved the use of The Pythagoras Theorem and finding the area of the right angled triangle. They were unable to use the area of the triangle to find the value of BX.

Concerning the responses of boys and girls for this question, their responses were significantly different. In fact, a p-value of 0.000 was obtained for the Mann-Whitney U test.

Table 4.22: Mean for boys and girls to Question 15

Mean for boys	Mean for girls
2.26	1.81

Question 16: If $x = 1.2 \times 10^2$ and $y = 4.8 \times 10^{-1}$, express the values of

(a) $x + y$

(b) $\dfrac{x}{y}$

in standard form.

Objective: To test how students manipulate numbers written in standard forms.

Table 4.23: Response to Question 16

Full marks	Mean	Standard deviation
7	2.80	2.74

Students performed badly here, with a mean score of 2.80. Full marks were scored by 18.6% of the students, while 37.2% scored zero. Many students who could tackle this problem found the value of x as 120, the value of y as 0.48 then computed x + y and x/y and finally converted the answers into standard form.

An analysis gender wise showed that the responses of boys and girls to this question were significantly different (p= 0.000 for the Mann-Whitney U test).

Table 4.24: Mean for boys and girls in Question 16

Mean for boys	Mean for girls
3.42	2.24

Question 17: **The internal dimensions of a box are 10 cm by 8 cm by 5 cm. What is the maximum number of cubes of side 2 cm that can be stacked in the box?**

Objective: To test the spatial ability of students.

Table 4.25: Response to Question 17

Full marks	Mean	Standard deviation
4	0.50	1.31

This question was amongst the most difficult in this questionnaire, with a mean score of 0.50. 12.0% of the students scored full marks while the vast majority (87.3%) scored zero.

The common mistake that could be noted was:

$$\text{Number of cubes} = \frac{\text{volume of cuboid}}{\text{volume of cube}}$$

$$= \frac{400}{8}$$

$$= 50.$$

Considering this question from a gender perspective, there was a significant difference between the responses of boys and girls (p= 0.000 for the Mann-Whitney U test). This outcome reflects other research which asserts that girls experience difficulties in spatial visualization. In fact, a study by Battista (1990) concluded that males scored significantly higher than females on spatial visualization, geometry achievement and geometric problem solving.

Table 4.26: Mean for boys and girls in Question 17

Mean for boys	Mean for girls
0.93	0.10

Question 18 (i): **What can you say about b if b = a + c and a + b + c = 18?**

Objective: To find how students can manipulate symbolic statements

Table 4.27: Response to Question 18(i)

Full marks	Mean	Standard deviation
2	0.68	0.90

Approximately 61% of the students were unable to solve this problem. Around 29% scored full marks, while the remaining attempted only half the problem, thus scoring only 1 mark. This shows that our students encounter problems with the manipulation of algebraic statements. In Mauritius students are exposed to algebra at the secondary level, the primary level being devoted to arithmetic, basic geometry and graphs. However, these students have been experiencing algebraic manipulations for at least three and a half years and they were expected handle such problems easily.

Considering this question from a gender perspective, there is a significant difference between the responses of boys and girls ($p = 0.153$ for the Mann-Whitney U test).

Table 4.28: Mean for boys and girls in Question 18(i)

Mean for boys	Mean for girls
0.74	0.62

Question 18 (ii): **In a certain school there are 4 times as many girls as boys. There are 200 girls in the school. How many boys are there?**

Objective: To test how students interpret and manipulate word problems involving algebra.

Table 4.29: Response to Question 18(ii)

Full marks	Mean	Standard deviation
4	3.21	1.59

Approximately 80% were able to solve this problem correctly while 19.4% scored zero marks. The common wrong answer that was obtained was "800" which showed that those students experienced difficulties in understanding the language used and reversed the relationship.

An analysis gender wise showed that the responses of boys and girls to this question were significantly different (p= 0.001 for the Mann-Whitney U test).

Table 4.30: Mean for boys and girls in Question 18(ii)

Mean for boys	Mean for girls
3.42	3.01

Question 19: **In a game a student has to be always more than 2 m from each corner of a hall. The dimensions of the hall are 15 m by 10 m. The rectangle ABCD shown below represents the floor space of the hall. Show, by shading the area of the floor space in the diagram, the region in which the student can move.**

Objective: To test the visual abilities of students through the concept of locus.

Table 4.31: Response to Question 19

Full marks	Mean	Standard deviation
4	0.75	1.26

This is another question in which most of the students have not performed well. 70.5% scored zero while only 2.5% scored full marks. Note that 17.1% of the students scored 3 marks which showed that they understood the problem but only missed the point that the student had to be <u>more than</u> 2m from each corner. The common mistake noted was a rectangle within the rectangle ABCD.

An analysis gender wise showed that the responses of boys and girls to this question were significantly different (p= 0.000 for the Mann-Whitney U test).

Table 4.32: Mean of boys and girls in Question 19

Mean for boys	Mean for girls
1.02	0.51

Question 20: **Five numbers are listed below:**

5/6, 8.3%, 0.825, 0.83, 33/40. Find

(a) which two are equal.

(b) which one is the smallest.

Objective: To test how students manipulate rational numbers

Table 4.33: Response to Question 20

Full marks	Mean	Standard deviation
3	1.37	1.30

Approximately 31% of the students scored full marks while 41.8% scored zero in this question. Arithmetic is a prerequisite for algebra and a very sound knowledge and mastery of it definitely helps in enhancing the learning of algebra and the other different components of mathematics.

A Mann-Whitney U test concerning the responses of boys and girls for this question gave a p-value of 0.000 which showed that there is significant gender difference in the responses of students.

Table 4.34: Mean of boys and girls in Question 20

Mean for boys	Mean for girls
1.68	1.08

Question 21(a): **In a group of 100 students, 80 study Mathematics and 55 study Economics. If**

ξ = **{students in the group},**
M = {Mathematics students} and
E = {Economics students},
find the least and greatest possible values of

(i)   n (M ∩ E)
(ii)   n (M ∪ E)′.

Objective: To test how students deal with sets.

Table 4.35: Response to Question 21(a)

Full marks	Mean	Standard deviation
6	1.99	1.99

Approximately 13% of the students scored full marks while 31.1 % scored zero in this question. Note that those students who completed the question correctly used a Venn diagram to represent the situations. This is a very important technique and is useful when dealing with "Survey Problems". This technique reduces the bulk of word information into symbolic form which can then be used and interpreted easily. An analysis of the responses showed that there is no significant difference in the way boys and girls answered this question (p= 0.082 for the Mann-Whitney U test)

Table 4.36: Mean of boys and girls in Question 21(a)

Mean for boys	Mean for girls
2.13	1.86

Question 21(b) : If ξ = {Triangles}
              I = {Isosceles triangles}
              E = {Equilateral triangles}
              B = {Obtuse–angled triangles}
              R = {Right-angled triangles}
Draw a Venn diagram to illustrate these sets.
Comment on $(I \cup B \cup R)'$.
Objective: To test how students can represent the different types of triangles on a Venn diagram.

Table 4.37: Response to Question 21(b)

Full marks	Mean	Standard deviation
8	1.14	1.75

Approximately 60% of students scored a zero and 20.6% scored 2 marks. This question was a difficult one for majority of the students, showing a lack of conceptual understanding in a topic which they have been dealing with since the beginning of their secondary schooling (isosceles, equilateral and right-angled triangles are studied at primary level). Most of the questions set in this area are routine problems where students are asked to solve questions involving triangles using trigonometry or associated concepts. They are not used to items where the actual properties of the triangles and their interrelationship are tested.

An analysis of the responses of the students revealed a significant difference in the way boys and girls tackled this question ($p = 0.003$ for the Mann-Whitney U test).

Table 4.38: Mean of boys and girls in Question 21(b)

Mean for boys	Mean for girls
1.42	0.88

Question 22: **There are 12 red balls and x white balls in a bag. A ball is drawn at random. If the probability that a white ball is drawn is $\frac{2}{3}$, find the value of x. How many red balls should be removed from the bag so that the probability that a red ball is chosen is $\frac{1}{5}$?**

Objective: To test how students deal with probability.

Table 4.39: Response to Question 22

Full marks	Mean	Standard deviation
7	1.34	2.28

This is another question where the majority of the students encountered difficulty. Approximately 11% scored the full marks while 68.7% scored zero marks. Note that 18.5% scored 3 marks which showed that one third of the number of students were able to answer at least the first part of the question. Most of them proceeded as follows:

$$\frac{x}{x+12} = \frac{2}{3}$$

$$3x = 2x + 24$$

$$x = 24.$$

However there were a few who used proportion as follows:

$$\frac{1}{3} \rightarrow 12$$

$1 \rightarrow 36$

Therefore, $x = 36 - 12$

$= 24.$

An analysis of the responses of the students revealed a significant difference in the way boys and girls tackled this question (p= 0.000 for the Mann-Whitney U test).

Table 4.40: Mean for boys and girls in Question 22

Mean for boys	Mean for girls
1.94	0.81

Question 23: **The diagram shows the first 3 triangles in a sequence of equilateral triangles of increasing size. Each is made from triangular tiles of side 1 cm which are either black or white.**

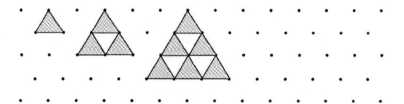

(a) **Draw the fourth equilateral triangle in the sequence on the diagram above, shading in the black tiles.**

(b) **Complete the following table for the equilateral triangles.**

Length of base (cm)	1	2	3	4	5	6
Number of white tiles	0		3			
Number of black tiles	1		6			
Total no of tiles	1		9			

(c) **Write down the special name given to the numbers in the last row.**

(d) **How many white tiles would there be in the equilateral triangle with base 10 cm?**

Objective: To test how students deal with patterns and deductions

Table 4.41: Response to Question 23

Full marks	Mean	Standard deviation
14	9.53	4.20

Approximately 21% of students scored full marks and 6.4% scored zero marks. This was one of the questions where almost all the students could attempt at least part of it. It should be noted that 26.5% of the students scored 12 marks out of 14 and these students could not find the name of the special numbers in the last row.

An analysis of the responses of the students revealed a significant difference in the way boys and girls tackled this question ($p= 0.000$ for the Mann-Whitney U test).

Table 4.42: Mean for boys and girls in Question 23

Mean for boys	Mean for girls
10.11	8.92

A summary of the analysis of the questions gender wise using Mann- Whitney U test is now provided in Table 4.43.

Table 4.43: Summary of results from Mann-Whiney U tests

Question	p-value	Significant Difference
1	0.152	No
2	0.854	No
3	0.001	Yes
4	0.026	Yes
5	0.158	No
6	0.260	No
7	0.002	Yes
8	0.269	No
9	0.000	Yes
10	0.706	No
11(i)	0.234	No
11(ii)	0.885	No
12	0.331	No
13	0.000	Yes
14	0.014	Yes
15	0.000	Yes
16	0.000	Yes
17	0.000	Yes
18(i)	0.153	No
18(ii)	0.001	Yes
19	0.000	Yes
20	0.000	Yes
21(a)	0.082	No

21(b)	0.003	Yes
22	0.000	Yes
23	0.000	Yes
Total	0.000	Yes
Number	0.000	Yes
Algebra	0.000	Yes
Geometry	0.000	Yes
Probability	0.000	Yes

Bivariate analysis

Chi square tests were conducted to establish possible association between certain factors and performance in the mathematics test. The grade in the test was found from the actual performance using the key described in Table 4.44. These were chosen after analyzing the overall performance of the students and the level of difficulty of the question paper.

Table 4.44: Range of marks for the grade in the test

Marks in the test (x)	Grade obtained
$x \geq 60$	A
$50 \leq x < 60$	B
$40 \leq x < 50$	C
$x < 40$	D

The first Chi square test involved the grade obtained in the test and the gender of the student. The result is as shown in Table 4.45 below:

Table 4.45: Gender and Grade

	Grade			
	A	B	C	D
Male	84	45	63	96
Female	43	56	63	156

The number of males scoring an A in the test is almost twice the number of females in that category, while number of females scoring a D is $1\frac{1}{2}$ times the number of males in that category. This suggests that, in general, boys performed relatively better in the test than the girls. A χ^2 value of 27.302 was obtained indicating that there is a significance difference between the performance of boys and girls in the test (p value = 0.000).

An analysis of the performance with respect to Ethnic Community was also performed. The contingency table was as follows.

Table 4.46: Ethnic Groups and Grade

	Grade			
	A	B	C	D
Hindu	55	59	75	135
Muslim	22	10	22	32
Chinese	30	9	1	2
General Population	19	21	28	77

A χ^2 value of 88.524 was obtained which revealed that there is an association between the grade obtained in the test and the ethnic community (p value = 0.000). The relationship between the performance in the mathematics test and ethnic community will be further dealt with while carrying out regression analysis in a later section.

An analysis of the performance in the mathematics test with respect to the "zones" was also carried out.

Table 4.47: Zones and Grade

	Grade			
	A	B	C	D
Zone 1	0	8	31	33
Zone 2	28	35	36	65
Zone 3	46	26	28	96
Zone 4	46	23	24	45
Rodrigues	7	9	8	13

A χ^2 value of 61.211 (with a p value 0.000) was obtained which showed that there is a significant difference between the mathematics achievements of the students in the test in the different zones. The zoning system for the admission of secondary students was recently established in Mauritius (since January 2003) meaning that a student is admitted to a secondary school in the zone where he/she resides. This analysis indicates that there is a significant difference in the performance of the students in the different zones.

Analyses regarding private tuition in mathematics at Form I, Form II, Form III and Form IV levels and the total grade obtained in the mathematics test administered were carried. The contingency tables for the four forms with respect to the grade in the mathematics test are given below.

Table 4.48: Private tuition in Form 1 and Grade

		Grade			
Private		A	B	C	D
Tuition in	Yes	9	8	11	24
Form 1	No	115	89	111	197

Table 4.49: Private tuition in Form 2 and Grade

		Grade			
Private		A	B	C	D
Tuition in	Yes	26	21	34	55
Form 2	No	100	77	90	171

Table 4.50: Private tuition in Form 3 and Grade

		Grade			
Private		A	B	C	D
Tuition in	Yes	66	46	75	119
Form 3	No	60	53	51	113

Table 4.51: Private tuition in Form 4 and Grade

		Grade			
Private		A	B	C	D
Tuition in	Yes	113	75	87	177
Form 4	No	14	26	35	70

χ^2 tests showed that private tuition at Form 1, Form II or From III levels did not have much influence in the total grade obtained in the mathematics test administered (χ^2 values 1.396, 1.934 and 4.075 respectively). Taking private tuition in mathematics is not popular in Form I and Form II in Mauritius (9.22% and 23.52% respectively), but by Form III this figure has risen to 52.49 %. The percentage of students taking private tuition in mathematics at Form IV level reaches 75.38 %. This is a common phenomenon in Mauritius. Private tuition exists right from the primary level, especially at the Standard V and Standard VI (end of primary cycle). This practice subsides in the lower secondary level to re-appear again as from Form III or Form IV level. However a χ^2 value of 15.958 has shown that there is an association between taking private tuition at Form IV level and the mathematics achievement in the test. The relationship between number of hours of study at home and performance in the mathematics test was also analysed. The contingency tables in this respect are as follows:

Table 4.52: Hours of study in Form 1 and Grade

		Grade			
		A	B	C	D
Hours for	2-4	67	71	82	172
home study	4-6	32	20	29	41
in Form 1	6-8	16	6	10	9
	More than 8	11	2	2	3

Table 4.53: Hours of study in Form 2 and Grade

		Grade			
		A	B	C	D
Hours for	2-4	54	50	61	139
home study	4-6	37	41	47	71
in Form 2	6-8	21	6	14	12
	More than 8	14	2	2	4

Table 4.54: Hours of study in Form 3 and Grade

		Grade			
		A	B	C	D
Hours for	2-4	26	32	39	89
home study	4-6	48	31	43	87
in Form 3	6-8	25	28	34	41
	More than 8	27	8	9	10

Table 4.55: Hours of study in Form 4 and Grade

		Grade			
		A	B	C	D
Hours for	2-4	13	20	23	66
home study	4-6	32	30	39	102
in Form 4	6-8	38	26	39	45
	More than 8	44	23	24	34

χ^2 tests proved that there was a relationship between the number of hours of study at home and performance in the mathematics test, as shown by the χ^2 values displayed below.

Table 4.56: χ^2 values for performance in the test and hours of study

Number of hours of home study	χ^2 values	Significance level
Form I	33.996	0.000
Form II	44.768	0.000
Form III	41.386	0.000
Form IV	44.721	0.000

The amount of time devoted to study at home is very important to review the concepts dealt in the classroom and consolidate the conceptual understanding. Moreover, tackling problems plays an important role in the learning of mathematics. The more time is devoted to mathematics the better tends to be the achievement in the subject.

The importance of prior performances of students in mathematics in classes in their achievement in the test was also analysed. It should be pointed out that the range (0-40, 41-60, 61-60 and more than 80) correspond to the marks obtained by the students in the sample in their final year school examination for each Form. The contingency tables are as follows:

Table 4.57: Performance in Form 1 and Grade

		Grade			
		A	B	C	D
Performance	0-40	0	2	14	41
in Form 1	41-60	6	35	43	108
	61-80	35	40	50	67
	More than 80	82	23	16	14

Table 4.58: Performance in Form 2 and Grade

		Grade			
		A	B	C	D
Performance	0-40	1	3	12	50
in Form 2	41-60	7	28	39	103
	61-80	26	49	55	69
	More than 80	90	20	16	8

Table 4.59: Performance in Form 3 and Grade

		Grade			
		A	B	C	D
Performance	0-40	0	5	12	52
in Form 3	41-60	5	16	21	105
	61-80	25	55	66	53
	More than 80	93	23	23	12

Table 4.60: Performance in Form 4 and Grade

		Grade			
		A	B	C	D
Performance	0-40	0	1	7	36

in Form 4	41-60	1	9	14	50
	61-80	8	19	32	25
	More than 80	52	18	9	7

It has been found that prior performances of students in the other classes had an influence in the performance in the mathematics test. These are shown by the χ^2 values below.

Table 4.61: χ^2 values for performance in the test and prior performance

Performance in	χ^2 value	Significance level
Form I	214.965	0.000
Form II	272.654	0.000
Form III	295.842	0.000
Form IV	171.384	0.000

Mathematics is a hierarchical subject and one has to master the prerequisite concepts and skills before moving onto other concepts. This view was verified through informal interviews with past students who have not performed well in mathematics at the School Certificate level. Many attributed their non-success in mathematics to some instances of failure, or problems experienced at the lower secondary level. Those who did remedy the situation could move forward, but those who could not do much to review the situation (or did not have enough opportunities provided) continued with this 'handicap' and consequently experienced further difficulties in mathematics later on

It can be noted that, from the contingency tables, that there are a number of students who performed well in their examination at Form III level but did not perform as well in the mathematics test administered for this study. This may be so because of the way the items were written. The questions were designed to test conceptual understanding. This analysis tends to suggest that the students may be at ease at answering routine questions with an algorithmic approach to problems, but have difficulties with conceptual understanding related to certain mathematical concepts. This finding is in line with a study conducted at the primary level in Mauritius (Monitoring Learning Achievement, 2003). It has been found that the majority of students involved in the study experienced problems in the items of mathematics that were written in an unfamiliar way.

The influence of the performances at the CPE level (end of primary national examinations) in the different core subjects on the performance in the mathematics test was also analysed. It should be mentioned that at this national examination, students are awarded grades by the Mauritius Examination Syndicate in each of the core subjects and not actual marks. The contingency tables for each core subject are shown below.

Table 4.62: Performance in English at CPE and Grade

		Grade			
		A	B	C	D
	A	122	71	74	121
CPE	B	2	15	25	57
English	C	0	10	13	27
	D	0	2	7	21
	E	1	1	6	19

Table 4.63: Performance in Mathematics at CPE and Grade

		Grade			
		A	B	C	D
	A	124	71	75	111
CPE	B	0	16	15	29
Mathematics	C	1	8	16	38
	D	0	3	11	38
	E	0	2	9	30

Table 4.64: Performance in French at CPE and Grade

		Grade			
		A	B	C	D
	A	123	75	81	151
CPE	B	1	9	18	32
French	C	0	8	16	32
	D	1	6	9	18
	E	0	2	2	10

Table 4.65: Performance in Environmental Studies at CPE and Grade

		Grade			
		A	B	C	D
	A	124	80	87	147
CPE	B	0	9	17	36
Environmental	C	1	7	9	21
Studies	D	0	2	9	28
	E	0	2	3	9

χ^2 values relating to the relationship of the performance in the core subjects at the CPE level and the performance in the mathematics test are given below.

Table 4.66: χ^2 values for performance in the test and performance at CPE

Subject at CPE level	χ^2 value	Significance level
English	94.957	0.000
Mathematics	124.105	0.000
French	62.216	0.000
Environmental Studies	70.002	0.000

This analysis shows that not only does the performance in mathematics at the primary level play an important role in how students achieve in mathematics at the secondary level, but also the performance in the other core subjects. Very often language is stated as a major difficulty for students in the learning of mathematics. At times students cannot solve a problem because they cannot interpret the message given in a word problem.

The relationship between the performance in the Oriental Language and that in the mathematics test was also analysed.

Table 4.67: Performance in Oriental Language at CPE and Grade

		Grade			
		A	B	C	D
	A	47	32	52	76
CPE	B	4	9	12	23
Oriental	C	6	8	8	22
Language	D	8	4	4	11
	E	8	7	9	17

A χ^2 value of 15.705 (with a p value of 0.402) showed that the performance in Oriental Language did not have a significant influence on the achievement of the students in the mathematics test. Note that oriental language was not taken by all students at the CPE level. The performance in this subject was not taken into account for the ranking purpose at the CPE examinations but was for grading the 2004 examinations.

The relationship between parents' education and profession with the performance in the mathematics test was also carried out. The contingency tables are as follows.

Table 4.68: Father Education and Grade

		Grade			
		A	B	C	D
	Primary	8	24	36	88
Father	Secondary	51	47	65	128
Education	University	37	18	16	17
	Professional Training	28	11	7	10

Table 4.69: Mother Education and Grade

		Grade			
		A	B	C	D
	Primary	11	30	35	87
Mother	Secondary	63	56	71	139
Education	University	31	10	8	8
	Professional Training	19	5	8	11

Table 4.70: Father Profession and Grade

		Grade			
		A	B	C	D
	Unemployed	1	1	2	7
Father	Manual/Unskilled	24	49	72	147
Profession	Clerical	13	10	15	35
	Middle Officers	34	16	20	31
	Managerial	21	7	7	8

Professional	24	10	3	4

Table 4.71: Mother Profession and Grade

		Grade			
		A	B	C	D
	Unemployed	57	61	79	134
Mother	Manual/Unskilled	6	12	27	61
Profession	Clerical	14	7	7	19
	Middle Officers	24	12	10	18
	Managerial	12	4	3	9
	Professional	8	2	1	1

χ^2 tests performed showed that all the factors relating to parents education and profession were significantly related to the performance in the mathematics test.

Table 4.72: χ^2 values for performance in the test and parents education and profession

	χ^2 value	Significance level
Father education	92.467	0.000
Mother education	80.548	0.000
Father's profession	109.566	0.000
Mother's profession	65.415	0.000

This finding is in line with other research highlighting the contribution of parents in the educational achievement of their children (Carr, Jessup, & Fuller, 1999; Kratsios & Fisher, 2003; Monitoring Learning Achievement, 2003). Parents spend a considerable amount of time with their children, helping them in their academic endeavors. The way parents support the children can motivate them, and can also provide logistic and other facilities to make a difference when it comes to the achievement. Parents who themselves have experienced difficulties in mathematics may convey negative messages to their children and this may, in some way or another, influence the attitude of the child towards the subject and eventually his/her performance in the subject. Research has shown that parental and societal attitudes play an important role in internalizing the feeling among girls that they are inferior to boys in mathematics (Campbell & Mandell, 1990; Eccles et al., 1987; Parsons, Kaczala, & Meece, 1982, cited in Casey, Nuttal & Pezaris,, 2001).

It was found that there is a significant difference between the performance of boys and girls in the mathematics test. The performance of boys and girls in mathematics at the CPE level was analysed. The contingency table was as follows.

Table 4.73: Grade in CPE mathematics and Gender

	CPE Mathematics Grade				
	A	B	C	D	E
Male	286	26	32	20	19
Female	310	34	31	32	22

A χ^2 value of 3.120 (with a p value of 0.538) showed that no difference was noted when the performance in mathematics at CPE (end of primary national examination) level were analysed with respect to gender. This is in keeping with research findings which assert that gender differences in mathematics are not found at primary level, but rather that they occur at the outset of adolescence (Fennema, 2000).

Note that in all the four strands, there is a significant gender difference in the performance of the students. This is in line with other research which has shown that the areas in which males are advantaged are those where "visualizations can be beneficial (e.g. geometry, measurement, proportional thinking, word problems, estimation)" (Johnson, 1984; Lummis & Stevenson, 1990; Marshall & Smith, 1987; Mullis et al., 1998; Robitaille, 1989; cited in Casey, Nuttal & Pezaris, 2001, p. 30).

A summary of the results obtained for the bivariate analysis is shown in Table 4.74.

Table 4.74: Summary of Chi Square tests conducted

First variable	Second variable	Pearson Chi-Square	Significant?
Total Grade	Sex	27.302	Yes
Total Grade	Ethnic Community	88.524	Yes
Total Grade	Zone	61.211	Yes
Total Grade	Private Tuition in FI	1.396	No
Total Grade	Private Tuition in FII	1.934	No

Total Grade	Private Tuition in FIII	4.075	No
Total Grade	Private Tuition in FIV	15.958	Yes
Total Grade	Hours of home study I	33.996	Yes
Total Grade	Hours of home study II	44.768	Yes
Total Grade	Hours of home study III	41.386	Yes
Total Grade	Hours of home study IV	44.721	Yes
Total Grade	Performance in Form I	214.965	Yes
Total Grade	Performance in Form II	272.654	Yes
Total Grade	Performance in Form III	295.842	Yes
Total Grade	Performance in Form IV	171.384	Yes
Total Grade	CPE English Grade	94.957	Yes
Total Grade	CPE Maths Grade	124.105	Yes
Total Grade	CPE French Grade	62.216	Yes
Total Grade	CPE EVS Grade	70.002	Yes
Total Grade	CPE Oriental Grade	15.705	No
Total Grade	Father Education	92.467	Yes
Total Grade	Mother Education	80.548	Yes
Total Grade	Father's profession	109.566	Yes
Total Grade	Mother's profession	65.415	Yes

An analysis of the relationship between attitude towards mathematics and the performance in the test was made. A correlation of 0.336 was obtained which is significant at the 0.01 level. A correlation coefficient (Spearman's Rho) of 0.449 was obtained concerning the aggregate at CPE level and the performance in the mathematics test. This correlation was found to be significant at the 0.01 level.

Regression analysis

The data was further analyzed using regression analysis and following regression equations were obtained:

Number score = 3.828 + 1.426 P4 + 0.831 P3 + 1.673 Chinese

Algebra score = − 0.435 + 7.586 P4 + 6.068 Chinese + 0.497 SES1 + 1.655 H1 + 2.30 NewSex

Geometry score = − 1.505 + 0.367 SES1 + 0.799 P4 + 1.638 NewSex + 2.537 Chinese + 1.699 Muslim − 5.598 NewRep2

Probability score = − 3.234 + 0.844 P4 + 0.218 SES1 + 1.221 NewSex + 1.751 Chinese + 0.498 P1

Total score = 1.290 + 11.341 P4 +1.308 SES1 +12.467 Chinese + 6.036 NewSex − 28.944 NewRep2

where

Number score = total score in the questions in the mathematics questionnaire from the Number strand

Algebra score = total score in the questions in the mathematics questionnaire from the Algebra strand

Geometry score = total score in the questions in the mathematics questionnaire from the Geometry strand

Probability score = total score in the questions in the mathematics questionnaire from the Probability and Statistics strand

Total score = total score in all the questions in the mathematics questionnaire

P1 denotes performance in Form 1

P3 denotes performance in Form 3

P4 denotes performance in Form 4

SES1 denotes the socio economic status index

H1 denotes the number of hours of home study in Form 1

NewSex = 1 if student is a male
        0 if student is a female

Chinese = 1 if ethnic community is Chinese
          0 otherwise

Muslim = 1 if ethnic community is Muslim
         0 otherwise

NewRep2 = 1 if student has repeated Form 2
          0 otherwise.

Note that gender does influence the performance in Algebra, Geometry, Probability and the total score but not in Number. The inclusion of NewSex with a positive coefficient in a regression equation shows "maleness" in that domain. Another important thing that can be noted from the regression analysis is the influence of culture in the performance in mathematics. In fact one can note the presence of the variable *Chinese* in all the regression equations which show that the students from the Chinese community tend to perform well in mathematics in general. This is in accordance with the findings of TIMSS studies where the performance of Asians countries in science and mathematics was outstanding, and also in studies conducted by Rao, Moely & Sachs (2000). These authors noted that "Chinese parents typically have high expectations for their children's academic success and perceptions of these expectations influence children's mathematics performance" (Rao, Moely & Sachs, 2000, p. 288).

It should however be pointed out that apart from the comment made above, little more can be said concerning the influence of culture in mathematics performance since the number of participants from each ethnic community in the sample were quite different as shown in the table below.

Table 4.75: Number of each Ethnic Community in the sample

Community	Number
Hindu	324
Chinese	42
Muslim	86
General Population	145

Also note that repeating Form 2 has been found to play an adverse role in the geometry score and the overall performance. Form 2 can be termed as a crucial stage in the educational life of a child at secondary level. In Form 1 the students have just joined secondary schools and many chapters are in fact repeating some ideas from the Standard Six syllabus to ascertain smooth transition from primary to secondary schooling together with some new concepts which are introduced at an appropriate pace. However, in Form 2, many of the foundational concepts needed for the secondary mathematics are dealt with and if students have not mastered these properly (and have not remedied the situation) they tend to encounter problems in subsequent classes.

Whenever the 'attitudetotal' (score in the attitude questionnaire) was added for the regression analysis, the following equation was obtained:

Total score = − 2.065 + 11.521 P4 + 1.239 SES1 + 11.364 Chinese + 6.005 NewSex − 27.030 NewRep2 + 2.223 H1.

It is to be noted that 'attitudetotal' does not appear in the regression equation, and the contribution of the other variables are almost the same (the coefficient of each variable in the two regression equations for total score are almost the same), but a new variable has now appeared, namely H1 (number of hours of home study in Form 1) and the constant term is now negative.

ANOVA results

ANOVA tests were carried out to find a relationship between ethnic community and the performance in the different strands of the test and also with attitude towards mathematics:

Table 4.76: Results of ANOVA tests

Strand	F value	Significance
Number	12.927	0.000
Algebra	27.715	0.000
Geometry	31.952	0.000
Probability	27.778	0.000
Total	33.692	0.000
Attitude total	1.672	0.172

Again, it is to be noted that in all the strands, as well as the overall performance in the mathematics test, there has been a significant difference between the different communities. However, there is no significant difference in attitude towards mathematics in the different ethnic communities.

Mann-Whitney U tests were carried out to determine any significant gender difference in the performance of the students in the different strands, overall performance in the test and attitude towards mathematics.

Table 4.77: Mann-Whitney U tests for gender difference in performances

Strand	Z value	Significance
Number	− 3.951	0.000
Algebra	− 4.073	0.000
Geometry	− 6.656	0.000
Probability	− 4.315	0.000
Total	− 4. 856	0.000
Attitude total	− 1. 970	0.049

There was a significant gender difference in all the strands used for the test as well as in the overall performance. However, concerning attitude towards mathematics, the result was borderline significant.

Summary of the chapter

This chapter which deals with quantitative analysis has provided the first part of the analysis of data collected for this study. The responses of students to each item in the mathematics questionnaire designed for this study were analysed using the SPSS package and possible significant gender differences in these responses were investigated. It was found that boys performed significantly better than girls overall, and also in the Strands Number, Algebra, Geometry and Probability. Bivariate analyses were carried out, where chi square tests were performed. Associations of performance in the test with several background factors, like sex, Ethnic Community, education of parents and profession of parents, were established. The data were further analysed through regression analysis. Through the regression equations obtained it was found that sex of the child and the ethnic community, amongst other factors, were having an influence in the performance of the student in the different strand and overall in the test.

The analysis of data corresponding to phase two, which is mostly devoted to qualitative analysis, will be described in the next chapter.

CHAPTER FIVE
Analysis 2

After having obtained an overall picture of the performance of boys and girls in mathematics across Mauritius, a more in-depth analysis was carried out in Phase Two with a sample of four schools. In this chapter the findings of this second phase of the study are discussed. The sample consisted of four Form Four classes, i.e. one boys only school, one girls only school and two coeducational schools. Classroom observations were carried in these classes over a period of four months and the students completed the mathematics questionnaire which had been administered in the first phase. These completed questionnaires were checked by me and the students' responses to the items were coded for further analysis using the SPSS software. The Modified Fennema Sherman Mathematics Attitude Scale questionnaires were also administered to the students to measure their attitude towards this subject and these responses were also recorded using the SPSS package. To gain an insight into the relationship between the students and the teachers, the Questionnaire on Teacher Interaction (QTI) was administered to the students. The students were also asked to draw a mathematician and write down two reasons why they might require the services of a mathematician. This activity was designed to find out what images the students had of mathematicians. Interviews were also conducted with students, teachers, parents and other key informants. The outcomes of the use of all these different instruments for the second phase of the study are discussed in this chapter.

Mathematics Questionnaire

The design of this questionnaire was described in Chapter Two and it was used in Phase One. This questionnaire was administered again in the four schools chosen for Phase Two, and the resulting statistics concerning the Mathematics Question Paper which is part of the Mathematics Questionnaire are recorded in Table 5.1.

Table 5.1: Mathematics Question Paper Phase 2 — Mean and standard deviation in the different domains

Domain		Mean	Standard deviation
Number	**Boys**	**10.23**	**3.11**
	Girls	**9.74**	**3.19**
Geometry	**Boys**	**4.87**	**2.32**
	Girls	**3.61**	**2.53**
Probability	**Boys**	**2.39**	**2.29**
	Girls	**1.46**	**1.75**
Algebra	**Boys**	**24.53**	**7.12**
	Girls	**21.17**	**6.72**
Total	**Boys**	**41.69**	**10.02**
	Girls	**35.05**	**10.43**

Note that in each of the domains, and in the total also, the mean performance of boys was higher than the mean performance of girls. This corresponds with the results obtained in the first phase of the study. To find out whether these differences were statistically significant, the Mann Whitney U test was carried for each of the domains and the results are presented in the Table 5.2.

Table 5.2: Mann-Whitney U tests for each of the domains

Domain	Z value	significance
Number	**-0.841**	**0.400**
Geometry	**-3.082**	**0.002**
Probability	**-2.243**	**0.025**
Algebra	**-2.484**	**0.013**
Total	**-3.265**	**0.001**

Tables 5.1 and 5.2 show that, other than the domain of Number, the difference between the performance of boys and girls in the other domains and in the total is statistically significant. This suggests that the performance of boys and girls was significantly different in mathematics. However, no significant difference was obtained concerning the attitude of boys and girls towards mathematics (p value of 0. 822 for the Mann-Whitney U test).

Questionnaire on Teacher Interaction (QTI)

The Questionnaire on Teacher Interaction has been extensively used in different parts of the world and its validity and reliability have been established: In the Netherlands (Wubbels, 1993), in Singapore (Goh & Fraser, 1995) and in Brunei (Ria, Fraser, & Rickards, 1997; Rickards, Ria, & Fraser, 1997). Recently it has been used in Mauritius at a tertiary institution (Bessoondyal & Fisher, 2003). In this present study it was used in each of the four sample classes. The findings for each teacher are presented in a tabular form showing the mean, standard deviation and Cronbach reliability coefficients for each of the scales involved in the QTI. A line graph of the means of the scales for all the four schools also has been drawn (Fig.5.5, p. 124) for comparison purposes.

Teacher of School A (Coeducational School)

This male teacher had 25 years of experience in secondary teaching. He had studied courses on pedagogy at the Mauritius Institute of Education leading to a Teacher's Diploma. After analyzing the responses of the students it was noted that this teacher had been rated highly on the Leadership, Understanding and Helpful/Friendly scales. However the results on the Student Responsibility scale implied that he was not providing sufficient opportunities for students to demonstrate this characteristic. His instruction tended to be teacher-centered, and students were listening to him most of the time. The means, standard deviations and Cronbach reliability coefficients for each of the QTI scales for this teacher are provided in Table 5.3.

Table 5.3: Mean and standard deviation for QTI scales for School A

Scale	Mean	Standard Deviation
Leadership	3.34	0.41
Understanding	3.39	0.30
Uncertain	0.69	0.36
Admonishing	0.91	0.38
Helpful/Friendly	3.38	0.24
Student Responsibility	1.58	0.87
Dissatisfied	1.10	0.23
Strict	1.89	0.83

To represent the characteristics of the teacher concerning the QTI scales visually, Figure. 5.1 has been drawn. This model has been used in many studies related to the use of QTI (Fisher, Rickards, & Fraser, 1996; Rickards, den Brok, & Fisher, 2003). Note that each shaded area represents measure of a particular behaviour. The further from the center of the chart a sector is shaded, the more prominent is the perception of the students about behaviour related to that scale.

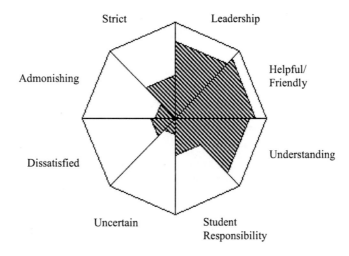

Figure. 5.1: Visual representation of QTI for teacher in School A

Note that the perception that the students have of this teacher is positive. The environment he creates should provide an atmosphere which is conducive to teaching and learning.

Teacher in School B (Coeducational school)

This male teacher had 19 years of teaching experience. He completed his Post Graduate Certificate in Education from the Mauritius Institute of Education some ten years earlier. The analysis of the responses of students to the QTI revealed that the mean scores on each of the scales were almost the identical – near the average except for the Admonishing scale, for which the mean score was the highest (shown in Table 5.4).This showed that the teacher was not performing well on several important qualities. Most likely, this was to the detriment of an appropriate environment for enhancing teaching and learning.

Table 5.4: Mean and standard deviation for QTI scales for School B

Scale	Mean	Standard Deviation
Leadership	1.97	0.39
Understanding	1.77	0.53
Uncertain	1.99	0.79
Admonishing	3.10	0.26
Helpful/Friendly	2.13	0.39
Student Responsibility	1.80	0.31
Dissatisfied	2.07	0.25
Strict	1.58	0.69

The visual model drawn in Figure. 5.2 clearly shows the difference in the various scales of QTI between this teacher and the one in school A.

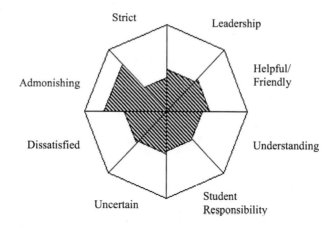

Figure. 5.2: Visual representation of QTI for teacher in School B

As can be observed from the diagram, most of the sectors are shaded up to 50% to the maximum except the sector which corresponds to Admonishing. Student concerns were also revealed from the interviews, which will be discussed later in this chapter, and this environment appeared to hamper the teaching and learning process.

Teacher of School C (Single Girls School)

This female teacher had 7 years of experience and had just completed her Post Graduate Course in Education (2004) from the Mauritius Institute of Education. Note that students' perceptions of this teacher were positive, as the mean ratings in Leadership, Understanding and Helpful/Friendly scales were quite high. However the teacher was viewed as being strict, and this view was supported by the value of the mean for the Strict Scale. The mean and standard deviation for each of the scales of the QTI for this teacher are shown in Table 5.5.

Table 5.5: Mean and standard deviation for QTI scales for School C

Scale	Mean	Standard Deviation
Leadership	3.54	0.23
Understanding	3.31	0.37
Uncertain	0.46	0.21
Admonishing	1.27	0.75
Helpful/Friendly	3.44	0.20
Student Responsibility	1.21	0.81
Dissatisfied	1.05	0.44
Strict	2.33	0.47

The pictorial representation of these data on the visual model is drawn in Figure. 5.3.

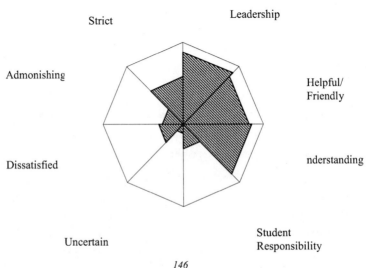

One can note the sectors corresponding to Leadership, Understanding and Helpful/Friendly are shaded to a great extent showing that the students believe that this teacher exhibits the behaviour described in those scales. It should however be mentioned that she was also seen to be exhibiting signs of strictness and this is further discussed when analyzing the interviews of the students.

Teacher at School D (Single Boys School)

This male teacher had 27 years of teaching experience at secondary level. He held a Teacher's Diploma from the Mauritius Institute of Education and was seen by his students to possess good leadership qualities, was confident in his work and was helpful. However, he was also seen to be strict and also he did not fare very well on the Students' Responsibility scale of the QTI.

Table 5.6: Mean and standard deviation for QTI scales for School D

Scale	Mean	Standard Deviation
Leadership	3.73	0.21
Understanding	3.58	0.30
Uncertain	0.43	0.50
Admonishing	0.74	0.45
Helpful/Friendly	3.49	0.45
Student Responsibility	1.44	0.87
Dissatisfied	0.57	0.27
Strict	2.47	1.05

Figure 5.4 below shows the data for this teacher in a pictorial way.

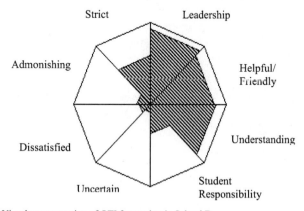

Figure 5.4: Visual representation of QTI for teacher in School D

147

Dissatisfied   Understanding

                                    Student
                Uncertain   Responsibility

Figure 5.4: Visual representation of QTI for teacher in School D

Here also the students perceived the teacher to be exhibiting good signs of
Leadership, Understanding and Helpful/Friendly behaviour, but they also described
the teacher as being Strict. The students' perceptions are examined more deeply at a
later stage in this chapter.

The means for each scale and for each of the four schools are now presented together
for comparison in a line graph. Note that while the three schools A, C and D are
comparable in each of the scales, School B stands out to be different in each of the
eight scales. This is more apparent in the following line graph.

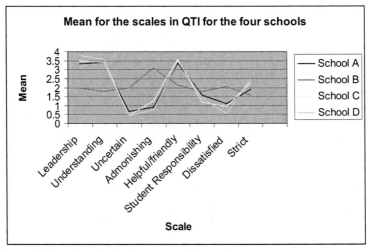

Figure 5.5: Line chart showing the means of all four schools together

The mean, standard deviation and Cronbach reliability coefficient for each scale when the data from the four schools are combined are given in Table 5.7.

Table 5.7: Mean, standard deviation and reliability coefficients of QTI scale

Scale	Mean	Standard Deviation	Cronbach reliability coefficient
Leadership	3.15	0.18	0.84
Understanding	3.03	0.19	0.80
Uncertain	0.90	0.33	0.76
Admonishing	1.49	0.30	0.83
Helpful/Friendly	3.10	0.20	0.83
Student Responsibility	1.52	0.58	0.60
Dissatisfied	1.17	0.21	0.74
Strict	2.08	0.73	0.70

Note that in general the perceptions of the students are positive, as the mean score in Leadership, Understanding and Helpful friendly are all above average while the score on Uncertain is quite low. It should also be noted teachers are to some extent viewed as possessing Admonishing characteristics and strict. Not many opportunities are normally given to students in mathematics classes to take responsibility for their learning and it seems that teachers tend to be the ones mostly involved in the classroom activities.

The next section describes the classroom observations that were carried out in the four classes identified for the second phase.

Classroom Observations

Classroom visits were carried out for a period of four months. Out of the four classrooms, three were in concrete, well furnished and the seats were arranged in rows pairwise. However, the fourth one was not properly built and the separations between the adjacent classes were with hardboard. One could hear whatever was being explained in the adjacent class. None of the classes had posters or any material relevant to the teaching and learning of mathematics on the walls. The seating arrangement was such that it was rare to see boys sitting next to girls in the two coeducational classes.

In all the four schools visited, instruction was teacher-centered, frontal and authoritarian. It was the teacher who was the main actor in the classroom. He/she was seen to be the main provider of knowledge and the one who took almost all the initiatives concerning the activities and happenings in the classroom. Students were quite passive and their main roles were to respond to the questions set by the teacher, or try problems that were assigned by the teacher. The questions that were set by the teacher were mostly at the Knowledge level, and sometimes at the Comprehension level. Few questions were asked by the students except occasionally they asked for clarification. The normal procedure was for the teacher to introduce a topic, discuss a few examples on the board on that topic which the students noted in their copybooks, and then assign some work for the students to try as classwork and homework. In case of doubts and questions from the students, the teacher intervened and provided some help. It was noted sometimes that students who were experiencing difficulties did not call out for the teacher but discussed the matter with their immediate friend.

Whenever homework were being corrected, one of the teaching strategies that was used was to send students to the board. Normally volunteers were sent to the board, and in very rare cases random selection of students were made. It was noted in the all girls' school that whenever a student could not attempt a problem, no appropriate prompts or help were given by the teacher — rather the teacher would instead solve the problem completely herself. However, in the all boys' school, whenever a boy who has been sent at the board had some difficulty in solving a problem, he was asked further questions to help him find his way. At times questions were directed to the other students in the class to initiate a discussion. The teacher sometimes did change the conditions given in the question and redirected the question to another student for him or her to try.

In one of the coeducational school, homework was normally corrected by the teacher orally with questions addressed to the class on the whole. There was usually a chorus answer, and once the teacher obtained the correct answer, he would proceed to the next question. There were a few random questions set to the pupils, but those who dominated in the class discussion often shouted the answer. In a few cases students selected randomly were sent to the board and the others were asked to comment on his/her answer, or to help him/her solve the problem. It was noted that the teacher interacted more regularly with the boys as they were the ones calling out more often, though there were a number of girls who would also make their voices heard. If there were questions which could not be answered by any student, the teacher would solve them completely on the board.

The strategy used by the teacher in the other coeducational school was somewhat different: each question was discussed with the participation of students, and whenever students worked on the board, they were sent in pairs: one boy and one girl. Both would solve the problem simultaneously on the board and discussions would follow on their method of work. One good point about this strategy is that both boys and girls were given practically equal teacher attention in the class. But one problem that the students may have felt was that their work was being compared in front of the class, and this may have caused some embarrassment to the student.

A copy of classroom observation schedule conducted in the schools appears in Appendix Four.

Student interviews

After having administered the different questionnaires to the students and having carried out classroom observations, a sample of nine students from each class were interviewed as a group for approximately one hour (see Appendix Five for a copy of the interview schedule). Arrangements were made with the school administration, and a room was made available. The students were chosen by the classroom teacher (three from the low achievers group, three from the average ones and three form the high achievers). The teacher was asked to use his/her marks in prior tests and his/her own judgment while choosing these students. The agreement of the students to participate was also sought. On the day of the meeting the objectives of the interview were explained to the students, and they were advised that all information collected from the interview was confidential and was to be used solely for the purpose of this study. The students were also asked to answer in any language in which they felt more at ease. In fact, all the students chose to answer in the home language "Creole" except for three students who initially answered in French and then switched to Creole. The response of the students to the different questions are analysed in the following section of this chapter. Whenever direct quotes are provided, a key is given to refer to the interview of the student in the appendix. For example *A6, SA II 3, p. 268* means Appendix Six, school A, the third student in Group II and which is found on page 268.

Do you find mathematics easy or difficult?

Students had mixed responses to this question. Those students who were at the lower end of the continuum (that is not performing well in mathematics) found the subject difficult. Some of them said that even after putting in considerable effort, they still found the subject difficult. Some of the direct quotations of students follow. Transcripts of all student responses to these items appear in Appendix Six.

- *"A bit difficult. I do not understand. Lots of confusion".* (A6, SA I 2, p. 268)

- *"A bit difficult. At times the teacher re explains but still we do not understand".* (A6, SC II 1, p. 279)

- *"Difficult. We need regular work in mathematics"* (A6, SA I 3, p. 268)

Thirteen students stated that mathematics was easy as well as difficult at different times. They claimed that it depended on the chapters under discussion and on the will of the student to work.

- *"A bit difficult and a bit easy. If we have understood all the explanations in the lower classes, it will be easy."* (A6, SA II 1, p. 268)

- *"It is easy as well as difficult. One has to make use of his head and reason out"* ((A6, SD I 2, p.283)

- *"Mathematics is a subject which is easy as well as difficult. There are people who do not have the ability to adapt with the subject. Some understand very fast while others do so after quite some attempts. It also depends on the chapters under discussion and on the way the teacher is explaining".* (A6, SB II 3, p. 271)

The students who were at the upper end of the continuum did report that mathematics was easy but they also said that effort was needed if they were to achieve well in the subject.

- "Mathematics is *easy and it is my favourite subject"* (A6, SA III 1, p. 268)
- *"Mathematics is easy if we practice regularly".* (A6, SD II 2, p. 283)
- *"I find mathematics very easy. One has to use his brains".* (A6, SD III 1, p. 283)

The boys were more confident in expressing themselves regarding mathematics being an easy subject. Certain girls, who were scoring well in tests and were rated as "good" by their teachers, had some hesitation in agreeing that mathematics was easy for them. The perception that a student has of mathematics influences his/her involvement and consequently his or her achievement in the subject. The more that one has a preference for the subject, the more motivated he/she will be to work and study the subject.

Do you find mathematics useful in everyday life? In what ways?

All the students who were interviewed considered that mathematics was useful. The most common uses given by them were: for shopping; the job market; its use in other subjects, and its value for further studies. One student also mentioned the use of mathematics in the development of reasoning powers. Students were aware of the importance that is attributed to mathematics for gaining employment. Without a reasonable grade scored in mathematics at the School Certificate level it was and still is difficult to secure a good job. Some of the direct quotations of students are given below:

- *"At each time we use it: mathematics is in fact about reasoning"* (A6, SA I 2, p. 268)

- *"It is really important as many jobs nowadays need mathematics"* (A6, SA II 3, p. 268)

- *"I do think that mathematics is very important, all the more for young children. Since childhood they have to have knowledge of mathematics. Their brain will function faster, they think fast and ahead. This will help them when they grow up – they will not take much time to take decisions"* (A6, SB III 1, p. 271)

How do you rate your performance in mathematics?

All the students were aware of their typical performance in mathematics and thirty of them felt that they should be able to perform better with more hard work. However there were a few (six in all) who admitted that they were having problems and were not progressing at a very fast rate. It should be noted that, in general, boys were more assertive about their ability to do mathematics than the girls. Some of the direct quotes are given below:

- *"I can say that I'm improving in mathematics. To be able to work well in mathematics, you not only have to understand the topic but also one has to practice. We have the great saying 'Practice makes perfect'."* (A6, SB II 2, p. 272)

- *"I am not satisfied. I think I should devote more time to maths as I am weak in it. The marks I score are quite low. I face difficulties while solving some problems"* (A6, SC I 2, p. 279)

- *"My performance in mathematics is low as compared to other subjects. I have to work harder and have more conviction. "* (A6, SD I 2, p. 283)

Do you encounter problems in mathematics or in particular topics? Which ones?

Only three of the students (2 boys and one girl) said that they did not have problems with mathematics and were quite at ease with the subject, while others mentioned particular areas of concern. Three students mentioned that on the whole things were fine except that they had problems with the language (the English one used in word problems). Several students (nine in all, mostly girls) mentioned that they were having problems in the topic **Circles**. They claimed that there were too many properties to remember. Other areas of concern mentioned were **Trigonometry**; **Three dimensions**, and **Locus**.

- *"Not as a whole but it depends on the topic. Where we have to remember formulae by heart and apply them"* (A6, SA I 1, p. 268)
- *"I do have major difficulties in mathematics as such for there are some topics where I have got some weak points. Like in Circles I do not know which theorem to use"* (A6, SB II 3, p. 273)
- *"In pyramids ... I have problems in calculations ... I have to understand the diagram to be able to proceed"* (A6, SC II 2, p. 279)

Has there been any particular event which has resulted in your liking/disliking of mathematics?

The point of this question was to find out whether something in particular occurred in the life of the student which resulted in him/her liking or disliking mathematics. Many students (almost all at School D) pointed out that they enjoyed mathematics because of the teacher they had in lower classes. This demonstrated the high esteem students had for their teachers, and also the great influence a teacher can have on students. However there were a few who mentioned that they did not like mathematics because of the teacher.

- *"I liked mathematics since the primary level. Whenever we were tackling mathematical problems amongst friends, it was so nice to be able to get the answer. There was competition but it was good. Also when we do well in assessment, this helps in developing a liking for the subject."* (A6, SB II 2, p. 273)

- "At primary level I was doing well in mathematics and also in Form I. However in Form II, I got a teacher who was explaining mathematics orally. We know that this subject cannot be done orally. I was facing much difficulty in that year and I did not do well in the exams either. This made me frustrated and I did not like mathematics at all. In Form III, I got a teacher who motivated us well and I started working well again" *(A6, SB III 1, p. 273)*

- *"I used to like mathematics a lot but in Form IV I am quite discouraged. I find mathematics becoming more and more difficult."* (A6, SC I 3, p. 280)

- *"I started to like mathematics since Form III because of my teacher. He motivated me and told me that I can do mathematics"* (A6, SD I 3, p. 284)

In which aspects of mathematics do you encounter problems?

The most common answer obtained from the students was that there were too many formulae to remember in mathematics and that often they were confused. They claimed mathematics calls a lot for the memory and this in itself is a cause of problems. Another major concern of students was the problem of language. They claimed that at times they knew the mathematical concepts but they found it difficult to understand the language in the problems. Learning through a second language is indeed a major issue in understanding mathematics.

- *"I think that the number of formulae we have to remember is too much. We may understand the problem but if we do not know which formula to use we shall not be able to do the problem. At times there are catching points in the questions"* (A6, SB II 2, p. 274)
- *"I think it is memory. It is quite difficult to remember all these formulae"* (A6, SB III 2, p. 274)
- *"One has to understand English properly to be able to do mathematics properly"* (A6, SD II 1, p. 284)
- *"Too many formulae. At times we have difficulties in understanding the questions"* (A6, SD III 3, p. 284)

What is the main factor that makes you like/dislike mathematics?

The objective of this question was to find out what students considered to be the main factor responsible for their liking or disliking of mathematics. Many students voiced the view that they liked mathematics because of the nature of the subject; it deals with reasoning and solving problems. Some liked to try hard problems, whereas others liked to 'play' with numbers. Here also the teacher was mentioned as the main factor responsible for the student's liking or disliking the subject.

- *"I like mathematics as I enjoy doing calculations."* (A6, SA III 1, p. 269)
- *"I like it because one has to strive hard to solve the problems,"* (A6, SA III 3, p. 269)
- *"If the teacher is friendly and can motivate us to learn the subject we will definitely do so."* (A6, SB III 2, p. 274)
- *"Myself. My readiness to work."* (A6, SD III 1, p. 285)

How do you rate the influence of your teacher in your achievement in mathematics? As already discussed, many students mentioned the contribution of their teacher in their learning of mathematics. This question was set to probe deeper into this issue. Students in two schools (one coeducational and the single boys) were unanimous about the positive influence of their mathematics teacher. The teacher in the coeducational school was rated high for his encouragement and accessibility to students. The teacher in the single boys' school was seen to be strict, but the students appreciated this strictness. They said that it helped them in studying mathematics. At the same time the patient character of this teacher was highlighted. On the other hand the teachers in the other two schools (coeducational and single girls) were reported by some students to be less encouraging: the way mathematics was taught in the classroom and the teachers' attitude towards the students were reported to be rather negative. It should be noted that girls were more prone to attribute their non-performance in mathematics to their classroom teacher while the boys had the tendency to associate their effort in mathematics to the classroom teacher.

- "My actual classroom teacher has brought in many things. In lower classes I was quite weak in maths but after working with him, my 'level' has increased." *(A6, SA I 1, p. 269)*

- *"For me the teacher explains very well. Whenever he is strict in class the students work better"* (A6, SD II 1, p. 285)

- *"She explains ok but at times she discourages us from learning mathematics"* (A6, SC I 2, p. 280)

- *"I think that the teacher is much more concerned with the completion of syllabus than helping students to understand mathematics. If we have not understood something he is reluctant to explain again. Moreover he corrects homework orally"* (A6, SB II 2, p. 275)

How do you rate the influence of your parents in your mathematics achievement?

The objective of this question was to find out the extent to which parents contributed to the learning of mathematics through their support. Almost all students, except for four (two girls and two boys), said that they did get support from their parents. What support existed varied from financial aid (paying for private tuition) to actually helping in the tackling of mathematics problems. Private tuition is very common in Mauritius at the upper secondary level, and mathematics is a subject where almost all students do take private tuition. It was found that parents encouraged students to realize the importance of mathematics for future education and also for the job market. There were instances where the support of the parents was due to the fact that the parents did not succeed in education themselves and thus they wanted their children to work hard and succeed. There also were cases where the student had been compelled to do well in mathematics because everybody in his family did well in the subject.

- *"My parents support me a lot. If I have to take tuition, they are ready to pay"* (A6, SA I 1, p. 269)
- *"If I need private tuition, they are there. But when it comes to help, my sister helps me as my parent work and have not time"* (A6, SA II 2, p.269)
- *"My parents encourage me a lot to study mathematics. They have got experience and they know how important mathematics is for a job"* (A6, SB I 2, p. 276)
- *"My father encourages me to do mathematics as he himself was good in it. He wants me to do even better. In Form II my performance in mathematics was not that good. He sent me for private tuition at two places"* (A6, SB III 3, p. 276)
- *"My parents encourage me to do mathematics. Almost everybody in my family do well in mathematics"* (A6, SD III 2, p. 285)

How do rate the influence of your friends in your mathematics achievement?
All 36 students who were interviewed agreed that their friends have an influence in their learning of mathematics. They stated that their friends help them in case of difficulties in mathematics, and there were also cases where the performance of other students in mathematics acted as a motivating factor for them to work harder. In fact, it was noted that girls were talking more about a cooperative environment in which they learnt while boys talked more about a competitive environment. The boys stated that this environment was conducive to learning and made them work harder.

- *"When there is something I haven't understood, my friends help me"* (A6, SA III 1, p. 270)
- *"Just like I help my friend, they also help me. We work together"* (A6, SD I 3, p. 286)
- *"When my friends score better marks than me, I tend to work harder to get better marks"* (A6, SB I 3, p. 276)

- *"Competition among friends helps us to work hard and do well in mathematics"* (A6, SB III 2, p. 276)

To what do you attribute your success/failure in mathematics?

Much research has been carried out in relation to attribution theory (Ernest, 1994; Fullarton, 1993; Kloosterman, 1990). The objective of this question was to find out if students were doing well or poorly in mathematics by their own admission and if they attributed their success or failure to any particular factors. If they were encountering problems, to what factors did they attribute this? It has been found through these interviews that doing poorly in mathematics was attributed by many students to a lack of effort and confidence, while success was attributed to the teacher, hard work or will to work. This was the case for both boys and girls.

- *"I think it is lack of confidence and practice. Not because it is too difficult for me as I know I can do mathematics"* (A6, SA I 1, p. 270)
- *"...Failure is because of my classroom teacher. He does not explain well"* (A6, SB I 1, p. 277)
- *"However well a teacher has explained, the effort must come from the student. They have to have the necessary motivation and work hard"* (A6, SB III 2, p. 277)
- *"I find mathematics a bit difficult. My teacher discourages me from doing mathematics. She tells me that the problems are difficult and I get discouraged"* (A6, SC I 3, p. 281)

Who has the greatest influence in your mathematics learning?

Eighteen students answered that those who had the greatest influence in their learning of mathematics were their parents. Out of these, most often it was the father who was mentioned. Some students did mention their friends. The teacher also was mentioned by thirteen students, but most often it was the private tuition teacher. As already stated, private tuition is very common in Mauritius at this stage of the secondary schooling. The private tuition teacher is chosen by the student or the parent, and the students will go to the tutor they are at ease and with whom they can interact. This may explain why many students mentioned their private tuition teacher to be the one having the greatest influence in their mathematics learning.

- *"My parents and in some way my private tuition teacher"* (A6, SB I 1, p.277)

- *"For me also it is my private tuition teacher. He is very friendly, explains well and is very helpful. Then come my parents and friends"* (A6, SB II 2, p. 277)

- "For me it is my friend. I am impressed by the way he deals with the problems. If he is not getting the right answer he keeps on trying hard" *(A6, SB I 2, p. 277)*

- *"My family. My father, mother, brothers and sisters have not done so well in education, they encourage me to do well in mathematics"* (A6, SA I 1, p. 270)

Do you prefer to have a male or a female mathematics teacher?

The objective of this question was to find out whether the sex of the mathematics teacher was influential on students as far as mathematics teaching and learning is concerned. Nine students stated that it is did not matter as long as the teacher explained things well. While no boy expressed a preference for a female teacher, there were some girls (five in all) who stated that they preferred a male teacher because he explained better or because their private tuition teacher was a male. It seems that girls have a small preference to work with male teachers.

- *"Does not matter provided he or she explains well."* (A6, SA I 2, p. 270)

- *"I think I shall be able to interact better with a male mathematics teacher provided he is elderly"* (A6, SB II 2, p. 278)
- *"If you carry a survey you'll find that most of the great mathematicians are males. Thus I believe that a male mathematics teacher will be better than a female one"* (A6, SB III 2, p. 278)
- *"Male teacher as my private tuition teacher is a male and he explains very well"* (A6, SC III 3, p. 282)

Do you find that mathematics is more for the male or female students?
The objective of this question was to find whether the students have a certain gender bias towards mathematics which may then affect their attitude towards the subject and consequently their involvement in it. While twenty four students stated that there is nothing of that sort operating, and that mathematics is for everybody, one boy and four girls did state that they felt that mathematics is more for boys, as girls prefer literature.

- *"Both have the same ability to learn"* (A6, SC II 2, p. 282)
- *"Boys are better in mathematics while girls are better in language"* (A6, SB II 1, p. 278)
- *"We cannot say for sure that boys are better or girls are better. It all depends on the student. However I'll agree that we girls are more prone towards literature which involves emotion and sentiments. However there are girls who are really good with numbers; we have exceptions"* (A6, SB II 2, p. 278)
- *"I do agree that boys have got a greater ability to do mathematics than girls. We girls are more attracted towards TV and thus language while boys are more interested with mechanical things"* (A6, SB III 1, p. 278)

Summary of the students' interviews
It can be noted, from these interviews, that teachers play a very important role in students' learning of mathematics. While the vast majority of students showed an appreciation of their classroom teacher, four of them asserted that they did not like mathematics because of their teacher. Teachers should be aware of their influence on

students and should ensure that no negative messages are sent to students. It also seems to be the case that prior achievement in mathematics has an influence on students' liking of the subject, and also on their confidence in mathematics. While recognising the help of peers and parents, students were of the opinion that their own effort was the most determining factor for improving their performance in mathematics. The students were fully aware of the importance of mathematics in their everyday life, further studies and in the employment sector. This, in itself, is a very strong basis on which the students' motivation to perform better in mathematics can be built by the teacher. Concerning the aspects of mathematics where students normally encounter problems, they were of the opinion that there were too many formulae involved and this could create confusion. The problem of language in word problems was also a cause of concern. The areas of difficulties for students in mathematics that were mentioned were Circles, Trigonometry, Three dimensions and Locus.

Interview of Teachers

In this section the responses of the four teachers in the sample to a semi structured interview conducted during the second phase are analysed. The interviews were held in the school the teacher was working in except for one, where arrangements were made to meet at a convenient place. The questions asked are written below first, followed by a brief analysis of the responses. Appropriate quotes are then given in italics with a key for reference to the Appendix. For instance, *A8, TB, p. 292* refers to Appendix Eight, teacher of school B to be found on Page 292.

How long have you been teaching mathematics at the secondary level?

Three of the teachers had teaching experience of 19 years or more. Only one of the teachers had around seven years experience. The number of years of service plays an important role in the professional life of the teacher. He/she comes to learn from the situations present in the classroom and also from the other staff members at the school, and these situations enrich their repertoire of teaching strategies and skills for classroom management.

Have you been teaching only boys, girls or both?

Two of the teachers (presently working in a co-educational school) had always worked in a co-educational system, while the teacher in the single-boys school had been involved only with boys, and the one from the single-girls school had been involved only with girls. It should however be noted that in private tuition, teachers work with both boys and girls at the same time.

Could you tell us more about your experience as a student of mathematics and as a mathematics teacher?

Three of the teachers were very fond of mathematics. They had liked the subject since their school days. They also enjoyed their job as a mathematics teacher.

- *"As a student of mathematics I have always been interested in the subject. Since primary school I have done well in mathematics. Even in secondary schools, I have won prize in mathematics."* (A8, TB, p. 292)
- *"Since my early years I liked mathematics. I was coming out first in mathematics always."* (A8, TD, p. 301)
- *"As a student I was very at ease with mathematics. My basic concept was to do mathematics step by step and logically. It was the subject I liked most."* (A8, TC, p. 296)

However one teacher acknowledged that he was not good at mathematics at school days. He became a mathematics teacher out of circumstances. But he is very happy with his job.

- *"As a student I did not like mathematics. But when I became a teacher I found mathematics was easier to teach than other subjects like physics."* (A8, TA, p. 289)

You have had pedagogical training at the MIE. How do you compare your teaching before and after the course?

All four teachers found their training to be of great help in their professional lives. They came to know and could practice different strategies.

- *"To be frank I have learnt much at MIE..."* (A8, TB, p. 292)
- *"My course was quite interesting. I had the opportunity of doing some research on teaching of mathematics and different strategies. It became part out parcel of my job..."* (A8, TD, p. 301)

However one teacher noted that she had problems while implementing the strategies learnt at the teacher training institution.

- *"This is where I had problems. After having followed the PGCE course, I wanted to put into practice some teaching strategies I learnt. I wanted to give them all the steps needed. They don't accept it."* (A8, TC, p. 297)

In my personal experience, these types of problems are quite common once a teacher finishes his/her professional training. Either they go back to their usual way of teaching or they are influenced by their friends at work to avoid innovative ideas. The Mauritius Institute of Education is trying to devise ways and means of supporting teachers at their school level even after their training is over.

What do you believe is the trend in the performance of students in mathematics in Mauritius?

Teachers all agreed that the general performance in mathematics is declining. Several reasons, such as lack of interest of students and the way of working in mathematics, were raised.

- *"I think we have a social problem. Students are not serious with their work. At our time we used to discuss mathematical problems with our brother, or anybody else. But the students nowadays do not come forward with their problems."* (A8, TB, p. 293)
- *"Students do not invest themselves too much in their studies. I believe that the performance could have been much better if they*

were better motivated. I believe that it is going downwards" (A8, TD, p. 301)

- *"The trend is that the students want the easy way out, very straight forward. They do not want to go through the basic concept of mathematics. They just want to know the answer and how to apply it. They just want to know the formula and then apply it. For example if you go through the concept of small change and limits, they do not want this. They just want to know the formula."* (A8, TC, p. 297)

What do you think are the difficulties that our students encounter while learning mathematics at the secondary level?

Concerning the main difficulties encountered by students in mathematics, the problem of language has been raised. Teachers are of the opinion that language presents a barrier to the proper learning of mathematics. They encounter difficulties while solving word problems.

- *"Regarding algebra, they may confuse the operations to use. In transformations, most of the students have problems. In the School Certificate paper there is choice. Students tend to neglect certain topics that they will not attempt in the examinations. Some students will tell you that they do not like this topic and do not like attempting these questions... I also feel that since children take private tuition, they do not concentrate in class."* (A8, TB, p. 293)
- *"Language is becoming more and more of a problem. The questions set by Cambridge need to be interpreted. Students cannot understand things, there is the language problem. Also there is the problem of calculations. Nowadays, students cannot handle figures. Student cannot do minor sums without the use of calculators."* (A8, TD, p. 302)
- *"Mental calculations. Students rely too much on calculators.* (A8, TC, p. 297)

- *"They have problems with their English language."* (A8, TA, p. 290)

Is there anything particular to boys or girls?

Two of the teachers were of the opinion that boys and girls do differ when it comes to the learning of mathematics. It was stated that boys appear to need less practice than girls to gain confidence, while girls practice much more. The attitude of girls towards the subject was also commented on.

- *"We can say that girls are more hardworking than boys. Girls come to seek help from the teacher more often than boys. Boys tend to work on their own or discuss with their friends. Some of the boys just try 4 or 5 questions on a topic and feel confident in that area but the girls no. Normally they do all the problems that you give... For the boys, they do easily grasp certain topics, even the difficult ones. But for the girls it is not that easy. They have to do more problems to be confident with the topic... I can say that boys perform better than girls and I think it is because of their ability."* (A8, TB, p. 293)
- *"Particular in the sense that girls have already put in their brains that mathematics is not for them. Mathematics is for the boys. I have been talking to the girls of Form I and they were telling me "Madam, we can't do mathematics. The boys can do it". I think they have got this impression right form primary level. Most of the teachers they meet at the upper primary level are males. Whenever they find a female mathematics teacher at secondary level, they are in a way astonished. For them a mathematics teacher is a male. They also make statements like "I hate mathematics" or "I cannot do mathematics".* (A8, TC, p. 297)

One of the teachers mentioned that very often girls have a negative perception of the subject right from primary days. They come to the secondary schools with that negative impression and consequently make no effort to improve their performance in mathematics.

- *"They join the secondary level with a negative impression towards mathematics and we have a lot to do to give them confidence. I also feel that at times it is the parents who convey this negative feeling towards mathematics. At times when we meet parents they tend to say*
    - *She is weak and has a poor memory*
    - *She will not be able to do mathematics Madam*

    We also have situations where students in front of the parents say:
    - *Oh Maths!. I can't do mathematics*

    The mother will then add:
    - *Even I could not do mathematics. So it is not much of a problem".* (A8, TC, p. 298)
- *I feel that teaching mathematics to boys is easier than teaching girls. It seems that boys tend to grasp the things faster than the girls. However when it comes to memory I find that girls do better. They follow the pattern set up by the teacher while boys tend to work fast and is more concerned with the answer."* (A8, TC, p. 298)

What role do you feel questioning plays in the teaching/learning process?

All the teachers were unanimous in recognizing the importance and use of the questioning technique in the teaching and learning process. They claim that its importance is felt more in a mathematics classroom. It is almost impossible to imagine a mathematics classroom where the questioning technique is not being used. However, it should be used judiciously and to help in the teaching leaning process.

- *"That's the most vital part of teaching. It is through questioning that we can obtain feedback from students."* (A8, TB, p. 294)
- *"Enormously. I just cannot do my job without questioning."* (A8, TD, p. 302)

However, one teacher did point out that it is not always easy to use that technique in class. There are instances when responses from students are almost nil and the teacher should find ways and means to ensure that interaction takes place in the classroom.

- *"Questioning helps the teacher to know whether the students have learnt the concepts and what they have learnt. However, students do not like to be questioned. They prefer the passive way of teaching. The teacher has much to do to ascertain interaction within the class."* (A8, TC, p. 298)

What can you say about the mathematical problem-solving skills of boys and girls?
While two of the teachers did not find much difference in the mathematical problem-solving skills of boys and girls, another from a co-educational school reported that girls tended to use the same strategy that was taught in the classroom while boys may opt for others.

- *"For boys they can choose any pattern to work while girls tend to use the pattern that was taught by the teacher. Boys can work many things mentally while girls have to write down everything on paper."* (A8, TB, p. 294)
- *"Girls tend to tell themselves that handling of instruments is not meant for them; I am thinking of locus problems; geometrical problems. But once you get them to overcome that problem they perform as well if not better than the boys."* (A8, TD, p. 302)
- *"I think the thinking process is the same. Previously girls were more methodical and presented their work stepwise while boy were more direct. But now they are almost the same."* (A8, TC, p. 298)

If you were offered a choice to teach mathematics to a single boys' class, or a single girls' class or a mixed one which one you would have preferred and why?

There were three different answers to this question: one teacher opting for single boys, two for single girls and one for a co educational situation. The two teachers from coeducational schools chose single girls, the main reason for that being the disciplined way of girls in a class. The female teacher in the single girls' school preferred a single boys class, while the one in the single boys school was in favour of a coeducational class.

- *"I'll go for single boys. Being of opposite sex will result in boys following the class with more seriousness."* (A8, TC, p. 299)
- *"I would have preferred only girls. They are a bit more attentive. Boys are more turbulent."* (A8, TA, p. 290)
- *"I'll prefer to work with girls. In terms of discipline, the way the girls work, they do not disturb. What you teach you get the response. It is easy to work with girls. They follow what you ask them to do."* (A8, TB, p. 294)
- *"I would prefer the third option. I think that there can be an element of motivation for both boys and girls to have them together."* (A8, TD, p. 302)

Do you feel that peers have a great influence in a student learning of mathematics?
Teachers do realize the contribution of peers in the learning of mathematics. They acknowledge that the students communicate in their own language and this might facilitate understanding. However one teacher pointed out that it is easier for boys to interact with friends, while girls tend to form smaller groups with whom they can discuss problems.

- *"When you talk about boys, they do. Boys can discuss with anybody while girls operate in a much smaller group. Boys are more open than girls".* (A8, TB, p. 294)
- *"Yes, enormously, I think it is quite natural. When a youngster has a problem he will try to find a solution from his friends and not us. So the peer group does help."* (A8, TD, p. 302)

- *"Yes, that is very true. Peers do have an influence. I know cases where students do not opt for say Additional Mathematics [an optional subject] (though she is capable) just because her friend has not opted. Even if the teacher tries to convince her by talking about her abilities, she will not opt for it. There are cases where students who were advised not to go for Add Mathematics opted for it because of their friend.* (A8, TC, p.299)
- *"Very often the students discuss in their groups. When one has a difficulty, another friend explains in his or her own language."* (A8, TC, p. 299)
- *"It depends on the intellectual level of the student. Those who are intelligent will make good use of peers and discuss with them. Others will not even bother."* (A8, TA, p. 290)

What about the contribution of parents?

The contribution of parents has been underlined, but parents have less time to devote to the education of their children. This is so if both parents are involved in the job market and at times working overtime. One of the teachers was of the opinion that at times parents tend to impose things on their children.

- *"Parents do help at times."* (A8, TB, p. 294)
- *"I see that parents have less and less time to look at their kid's work. But it is important for them to enquire on their kid's work."* (A8, TD, p. 302)
- *"Parents do have an influence. I know of one case where one girl was quite weak and could not do mathematics properly. She did not want to choose Add Mathematics as an option but her father was insisting that she must choose it (because she was choosing the Science stream). The girl could not work at all and was even crying. She failed her Form IV. The management even called the parents. The mother did understand that the girl was unable to do Additional Mathematics but she could not do anything as the father was almost forcing it on the girl."* (A8, TC, p. 300)

- *"Parents have a great role to play. Nowadays, parents do not have much time to look after their children. Proper monitoring is not there."* (A8, TA, p. 291)

What can be done to enhance the teaching and learning of mathematics at the secondary level?
Teachers were conscious of the problems students were facing while learning mathematics. Different ways to deal with the problem were suggested ranging from the use of history of mathematics to motivate students, catering for the difficulties right from the primary level, to changes that should be made in the curriculum materials.

- *"The way mathematics is taught in our schools is by giving rules and techniques to solve problems. I believe that if we show to the students some of the places and ways have mathematics is applicable it will increase the interest of students. History of mathematics also will help."* (A8, TB, p. 294)
- *"That's a question I have been asking myself. There is less and less of interest on the part of the students. The question is how to get the students interested in the subject. Getting them to be involved in it through projects. Kids nowadays are like this; they want to do things where they got pleasure. There are so many things (other than studies) that kids can derive pleasure from. It we could do something to make them be more interested in mathematics; getting them to be more involved in the subject. Derive pleasure from it. They do not fancy this nowadays; give them work, correct the homework, they've done it well and they are happy from it. They need something else. Through projects, they can be involved in history of mathematics and be more interested to mathematics."* (A8, TD, p. 303)
- *"I think changes should take place at the primary level itself. It happens that teachers at that level tell the student that you cannot do mathematics. I have a niece who is in Standard III. She is*

having some problems with mathematics. When I discussed with
her she told me that her teacher has told her "No problem.
Mathematics is not for you". This thing is put in the mind of the
child. Also the method students tend to do problems at primary
level; that is the trial and error method. Some of them have the
tendency to use it even at secondary level." (A8, TC, p. 300)

- *"I think the approach should not be too abstract. Also more of*
  solved examples in the textbooks. More help and support from the
  parents will definitely help." (A8, TA, p. 291)

Do you feel that gender equity in mathematics education has been achieved in
Mauritius? If no, what do you think can be done to achieve it?
There were mixed responses to this question. Two teachers believed that it has
been achieved while others said "not yet". It should however be pointed out that
one of the teachers did mention that there are girls in the mathematics class who
cannot do much as far as mathematics is concerned. It was suggested that time
cannot be wasted with these students, and it is better to concentrate more on the
ones who are doing well. I believe such an attitude on the part of a teacher itself
demands scrutiny, as it reveals that the teacher's interactions with his or her
students do not promote gender equity.

- *"I don't think that it has been achieved. From my experience, I*
  feel that boys will do well in mathematics but there are certain
  girls who cannot do much. One cannot devote much time to these
  girls. Girls tend to move to streams like language, economics."
  (A8, TB, p. 295)
- *"It's all in our mind. There are certain beliefs which are very*
  much in the minds and these have to be looked after. For instance,
  mathematics is not for girls. I believe teachers and the society
  have got that feeling. Also, female mathematics teachers are not
  that easily accepted by students. (A8, TD, p. 303)

- *"I don't think so. Parents do not influence girls too much to do mathematics. We teachers at the classroom try to ascertain gender equity. But for the society it is not so."* (A8, TC, p. 300)
- *"I do not find any difference between boys and girls."* (A8, TA, p. 291)

Summary of the teachers' interviews

The teachers were unanimous in recognising the contribution of pedagogical training in their professional lives. As already mentioned, pedagogical training prior to joining the teaching profession at the secondary level is not compulsory in Mauritius, unlike the case for the primary level. Concerning the trend in the performance in mathematics, all the teachers interviewed were of the opinion that there was a decline. The lack of interest of students was offered as one of the main reasons for these. The problem that students tend to encounter while solving word problems was also mentioned. One of the teachers interviewed (female teacher working in a single girls' school) pointed that there are cases where girls enter the secondary school with a negative attitude towards mathematics. The teachers then have a hard task to motivate these students to learn mathematics. She was also of the opinion that at times these negative feelings towards mathematics were conveyed to the girls by their parents, especially those who themselves had negative experiences with mathematics at school. However, the positive contributions of parents as well as peers in the learning of mathematics were also pointed out. Teachers were of the opinion that to be able to enhance students' performance in mathematics, one has to arouse their interest in the subject first. Other methods of teaching mathematics where the students are motivated should be sought out.

Interview of Parents

In this section the response of parents to a semi structured interview are analysed. The format for the analysis will be the same — that is, the questions will appear with a brief account of the responses to that question. This will then be followed by direct quotes from some parents in italics together with the key. For example *A10, P6, p 311* refers to Appendix Ten, Parent 6 to be found on Page 311.

You have a son/daughter who is in Form IV. How do you think he/she is progressing in studies?

The parents were all aware of the performance of their children in their studies. While some admitted that the child was having difficulties, they had encouraging words for the effort that was being put in.

- *"She is working but not as well as her sister"* (A10, P1, p. 305)
- *"I won't say he is doing well but he has progressed"* (A10, P2, p. 306)
- *"She is doing her best but she has to improve."* (A10, P3, p. 308)
- *"Works well. There are certain subjects where she is weak; English, Business Studies ..."* (A10, P5, p. 310)
- *"Generally in her studies she is not that good. But she is doing her best. We cannot guide her as both my wife and myself have studied only up to CPE level. The things are different. I can see that she is improving day by day. I have to give her private tuition. Without private tuition we cannot do much"* (A10, P7, p. 312)

What do you know about his/her performance in mathematics?

Regarding mathematics, it was noted that the parents did have mixed feelings concerning their children's performance. While some found that their children had been facing difficulties in mathematics since primary level, others felt that more effort was needed, and most probably private tuition might help. It should be noted that the parents of boys were generally more confident when they were talking about the performance of their sons in mathematics than the parents of the girls. In fact one parent of a girl mentioned that the girl's brother who is two years younger to that girl helped her at times in her mathematics lessons.

- *"In Mathematics, she is working well."* (A10, P1, p. 305)
- *"He was quite weak in maths as compared to my two other children. But with coaching he has improved. I would definitely wish that he performs much better but I am satisfied with his performance"* (A10, P2, p. 306)

- *"Concerning mathematics, she is ok. At times she tells me that she has not completed the questions. She needs more practice."* (A10, P3, p. 308)
- *"In mathematics, she has been facing problems. I find that his brother who is in Form II helps her at times in mathematics."* (A10, P4, p. 309)
- *"In maths she has improved. She was a bit weak. But now she is scoring better marks. I am giving her tuition. Also she is devoting more time to studies at home."* (A10, P5, p. 310)
- *"Since primary level she has been having problems in maths."* (A10, P6, p. 311)
- *"In maths, she is doing quite well. At certain times she is doing well but in the other she finds problems."* (A10, P7, p. 312)

Education is free in Mauritius but there are other expenses associated with it. Please indicate how you manage with the financial implications.

Parents were unanimous in recognising that education in Mauritius has considerable financial implications for them. While there are no school fees to pay, they have much to do to meet the expenses regarding books, transport and most importantly, private tuition. They acknowledge that much effort is needed on their part to be able to meet the expenses.

- *"A lot of expenses. We have books, tuition. She obtains books as a help but we have to wait a lot. We have to spend a whole day at the office. In fact this year we had to wait till April to get the books. There are some books that we have not yet obtained. We have to make a lot of sacrifices to be able to meet all the expenses. It is not easy at all.* (A10, P1, p. 306)
- *"A lot of expenses. We have books, transport, private tuition. I did not have the opportunity to study. I did only till Form IV and that time the monthly fee was Rs35. We could not pay. That's why I want to do everything to help my children in their studies. But we have a lot of sacrifice to make. For instance today is Saturday (5*

o'clock in the evening) and I have just reached home. We are doing all these for our children. I had an old computer and I just took a new one." (A10, P2, p. 306)

- *"It is not easy at all. I have three children at the secondary level and things are very difficult. We have books, transport and other things. I used to work before but I've stopped because of my father who is ill. I pray God that he helps me to continue supporting them for the education as her father works at times. We are doing everything that we can, but we hope that they value these by working hard."* (A10, P4, p.309)

Are you able to help him/her in mathematics at home? In what ways?

The majority of the parents interviewed noted that they could not do much to help their children in mathematics as either they had not studied much, or the mathematics they were exposed to was very different to the mathematics their children were doing. For instance, some thirty years back the topic "SETS" was discussed at the upper secondary level but nowadays it forms part of the lower primary curriculum.

- *"I cannot do much as I have studied till CPE only. But her sister helps her. I had to stop school to enable my brother to continue with his studies."* (A10, P1, p. 305)
- *"Personally I cannot. The mathematics that they are doing is different from what we did. That is why I give them private tuition. But on and off I check their work."* (A10, P2, p. 306)
- *"No, I cannot as things have changed a lot; especially mathematics."* (A10, P4, p. 309)
- *"I have studied only till CPE. Her father also cannot do much as he works extra time to meet the expenses. She has to manage on her own. She is the elder one in the neighbourhood and children in the vicinity come to her for help."* (A10, P5, p. 310)
- *"We do not find ourselves up to level to help our children. It is difficult for us as the maths we did at our time is different from the one our children are doing."* (A10, P7, p. 312)

Tell us about your mathematical experience at school.

Many parents did not have a pleasant experience with mathematics during their school days. That experience may have, in a way, affected the attitude of their children towards mathematics. It may happen that this negative experience is conveyed to the children and consequently affects their attitude towards mathematics.

- *"Mathematics has been a nightmare for me. When the teacher was next to me, I could manage with the problems. But once he was not there, I used to have problems. At our times, we did not have private tuition, we had to manage on our own."* (A10, P3, p. 308)
- *"At my times it was called Arithmetic. I was quite good in that subject till Form II. But I had a break as I went aboard for two years. When I came back I joined school again but I had many problems. I could not cope with studies and also because of the financial problems. I stopped school and joined the construction job. I however worked hard there and I got higher and higher responsibilities. This is the reason why I want to provide my children with all the necessary facilities for their studies. I do not want them to experience any break."* (A10, P2, p. 306)

There were however some parents who asserted that they did like mathematics during their time at school.

- *"I did like mathematics at school days"* (A10, P4, p. 309)
- *"At our time, maths was easy. But when I see what they study in mathematics at the primary level, nowadays it has changed a lot. There are so many new topics."* (A10, P5, p. 310)

What kind of support do you offer your children in the learning of mathematics? (for example special rooms, computer, books...)

Parents were unanimous in asserting that they ensure that the children received the necessary support to facilitate their learning of mathematics. They acknowledged that it costs them, but they did manage. One of the promising supports that parents mentioned was the opportunity for their children to take private tuition. They believed that much could be done to improve the performance of the children through private tuition.

- *"I am here to give them all the necessary support. They are not asked to help in household matters or other work. They should concentrate only in their studies.* (A10, P2, p. 306)
- *"Concerning support at home, I try to give her the basic things she needs. Our house is quite small but she has got a corner to study. Both my daughters like to work on a small blackboard which they have constructed themselves."* (A10, P1, p. 305)
- *"Concerning support, we give them everything. They do not give any help in household affairs to be able to concentrate on the education. At our times, there was discrimination. Parents preferred to educate their sons than their daughters. Girls were to stay at home at the expense of the boys.* (A10, P3, p. 308)
- "We talk to our children and ask them to tell us what they need. We make it a point to buy whatever book they need or provide any support they need for their studies. They have their own room for personal study." *(A10, P7, p. 312)*

Would the support be different if the child was a boy rather than a girl (or vice versa)?

The parents were unanimous and adamant in asserting that the support would not have been different if the child had been of a different sex. They were of the opinion that all children should be treated alike, and that equal opportunities should be given to them all. They however recognized that the scene was different during their time at school, where differences in opportunities and support were to be noted between boys and girls.

- *"No, all children should be treated the same way."* (A10, P1, p. 305)
- *"No not all. Previously it used to be different for boys and girls as parents were 'investing' on their sons."* (A10, P2, p. 306)

- *"No, not at all. I did not study as my sister was already at school and we wanted to give her the opportunity to continue. We were having financial problems also."* (A10, P5, p. 310)
- *"No, both are the same. They both need the same support and attention."* (A10, P7, p. 312)

To what extent do you yourself find mathematics useful?

Mathematics was valued to a great extent by all the parents. They found it to be very useful in everyday life, in other fields and also for job purposes. These views coincided with those of the students when they were interviewed. It seemed that this positive view of the use of mathematics had been communicated at home.

- *"Very useful. We need mathematics everywhere. As you can see I am selling cakes in front of my house. I need maths everyday."* (A10, P1, p. 305)
- *"I find mathematics is a must. Be it electronic, mechanic, accounting; in any field we need mathematics. Even in the kitchen. I find that mathematics is the number one."* (A10, P2, p. 307)
- *"Mathematics is really important. If one does well in maths, he/she will generally do well. I think that in mathematics there is logic."* (A10, P3, p. 308)
- *"Mathematics is very important. For any job we need mathematics. Even in everyday life. I have difficulties to deal with everyday situations myself. I can see the problems. That's why I want my children to be quite at ease with mathematics."* (A10, P4, p. 309)
- *"Very important. One cannot get a good job without maths. For example, in a bank, supermarket, etc. We use mathematics very much in our everyday life, like shopping, managing our budget, etc"* (A10, P5, p. 310)
- *"This is the most important subject. One should use his brains. I myself, being a vegetable-seller, i use it everyday. I do not need a*

calculator for the sums. Everything is done mentally. Also to get a good job." (A10, P6, p.311)

Do you feel that friends have an influence in a student's learning of mathematics?

Parents believed that peers had an influence in their children's learning of mathematics. They noted that there was much discussion between them, sometimes by telephone, concerning problems related to mathematics. They were however cautious of whom the friends are, and very often advised their children in the choice of their friends.

- *"Yes, it does help. They do discuss when they have problems."* (A10, P6, p. 311)
- *"We have always taught them that they should be amidst those who are more intelligent than themselves. They can learn from them. We believe that we should be friends with our children. We can talk over things thus give proper advice."* (A10, P7, p. 312)
- *"Yes, they do help themselves. Cooperation is very important."* (A10, P3, p. 308)

What do you think about the practice of private tuition?

This has been a topic of much debate in the Mauritian society. Many discussions have been made on this issue and a number of studies conducted (Goodoory, 1985; Rajcoomar, 1985). Much importance is still attached to it, especially at the upper primary level and upper secondary level. All parents interviewed confirmed that their children did take private tuition, and almost all commented on its positive contribution to the teaching and learning process. However, they all agreed that it has many financial implications and the parents have much to do to meet the cost.

- *"Concerning tuition, I give it to her only in those subjects where she has problems. It does demand much financial involvement but I try to manage. I should say that my daughters are satisfied with what help and support I can give them. I doubt whether a boy would have been satisfied."* (A10, P1, p. 305)

- *"Private tuition is something that I think will be difficult to stop. We, parents, know that the teachers at school are doing their best but our students may have some problem. Because of that we give our children private tuition so that they can catch up and cope in the class. We cannot choose the teacher at school but concerning the private coach we can choose. We'll choose the best and who will help our children."* (A10, P2, p. 306)

- *"I think that it helps. But if there was no tuition, the classroom teacher would have been forced to perform. In the older days, only those children who had difficulties were taking tuition. But now it is almost a fashion. Students of Form I take tuition. I think of those parents who have financial problems. They have much to do to ascertain that their children can take private tuition."* (A10, P3, p. 308)

- *"Private tuition does help. But at times I feel that it is a handicap. The ways the teachers proceed differ. This may create confusion. I however feel that if the children are conscious and hardworking, there is no need for tuition. Also tuition does involve much finance."* (A10, P4, p. 309)

- *"It is tight to cope but it is important. There are certain things that she does not understand in the class. She may ask her tuition teacher. We have to make sure that our children get all the opportunities and facilities to learn as we ourselves did not get the chance. She is taking tuition only in 3 subjects as I have other children also. I have to provide tuition to them also. I think that when she goes to Form V I'll be giving her tuition in more subjects. It does cost a lot but we have to manage. I have never worked before but I had to take up a job recently to help my husband financially."* (A10, P5, p. 310)

Summary of parents' interviews

Parents were found to provide their full support and encouragement to the education of their children, irrespective of their gender. This situation, as asserted by the parents who were interviewed, was different long ago when girls were 'sacrificed' in favour of their brothers for schooling. Many times this choice was made because of financial constraints and in some cases family obligations. Parents did recognise the importance of free education but still felt that there were considerable financial implications on their part to educate their children. One of the main financial obligations of parents in Mauritius was private tuition. In general, parents believed that private tuition does enhance the mathematics achievement of their children.

Draw a Mathematician Test

To get an idea of the perception that the students have of mathematics and of a mathematician, the Draw a Mathematician Test (Picker & Berry, 2000) was administered to them. Selected students had been interviewed and their responses to different attitudes related to mathematics were noted, but there may have been instances where their inner feelings were not honestly reported. In view of capturing these inner feelings which are not easily expressible in words, all the students involved in the second phase of the study were asked to draw a mathematician and to write down two reasons for needing the services of a mathematician. The drawings made by the Mauritian students had many of the characteristics depicted in various studies conducted abroad, such as the recent work of Sumida (2002) in China, Indonesia, Korea, the Philippines, and Japan. The Mauritian students' drawings were assessed using the standard basic indicators (Kahle, 1989) in DAST (relevant to a mathematician) and some others drawn from the work of Picker and Berry (2000). The results are shown in Table 5.8 on Page 166.

Table 5.8: Number of Boys and Girls Who Drew Stereotypical Indicators in their Drawings

	Number
Glasses	35
Facial hair	19
Symbols of research (e.g. mathematical instruments)	7
Symbols of knowledge (e.g. books)	22
Mathematical formulae	9
Beard	29
Bald	7
Calculator	1

It should be noted that only eight students out of the 81 students in the sample drew female mathematicians; seven being girls and one a boy. This observation reinforced the idea that students perceive mathematics to be a masculine activity. Furthermore, many students (49 in all) tended to imagine their teacher when they created their drawings. An example is illustrated in Figure 5.6.

Figure 5.6: Mathematician as a teacher

Very often the formulae that were represented in students' drawings were those found on classroom blackboards, or on papers lying on the teachers' table, or were those that students were actually working with in classroom activities. It also was significant that the mathematical scribbling in the drawings ranged from simple mathematics (like 1+2= 3) to meaningless writings involving mathematical symbols and operations. It was found that 14 students drew mathematicians as queer persons. An example of this is shown in Figure 5.7.

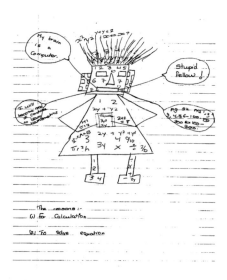

Figure 5.7: A mathematician as a queer person

In many cases the writings which accompanied the drawings give an insight to what the student thought about what mathematics is all about. A sample of the writings follows:

- *"Always angry, serious and in bad mood, hot tempered"*
- *"A mathematician looks very boring"*
- *"A mathematician becomes old very quickly because everywhere they go they want to do some calculation"*
- *"In my point of view the portrait a mathematician should be like this; that is by his appearance he should look wise and clever. He should be serious. He should look*

sophisticated and have a philosophical mind so that he can think in order to create maths formula"

- *'Stupid fellow"*
- *"My brain is a computer"*
- *"I will become mad if I continue to learn maths"*
- *"Wrinkles due to mathematical stress"*
- *"When a mathematics teacher has no hair on his head this means that he studies very hard to achieve something"*

Importantly, no drawing by any students in the Mauritian sample depicted a student or younger person. Picker and Berry (2000) reported that the drawings by students in almost every country chosen for their study showed any students drawn as small and powerless. In contrast, mathematicians were drawn as authoritarian and threatening.

When students were questioned about hiring the services of a mathematician, many answered that they would do this because it would help them in solving problems and force them to learn mathematics. The image of a mathematician as a personal tutor to help with or to solve complicated problems in mathematics can be found throughout the drawings and discussions with the students in this Mauritian study. It was revealed in the study that for students of this age, mathematicians in real world situations are too far removed from their imagination. The students relied on their teacher or stereotypical images from the media to provide images of mathematicians for the drawings produced in the study.

It was found throughout the drawing exercise that most students held a stereotyped image of mathematicians and mathematics. The drawings showed the mathematician as incompetent or having supernatural powers which suggested that the students had a negative attitude towards the subject or doubted their abilities in doing mathematics. These attitudes may be related to the invisibility of the mathematical processes in students' learning activities. The issue here is that when mathematical processes are not explicitly apparent for students, mathematics competence looks more like an external power, outside the real world of the student, rather than an ability which everyone has the potential to develop. It appeared that Mauritian students are influenced by their teacher role models and hold a very masculine image of the subject. It is a worrying indictment on mathematics in the school curriculum when students believe that learning mathematics has no benefits for them. One student (girl) wrote in her drawing:

- *"I think that mathematicians are needed to add problems to the lives of we people. They complicate our lives with their difficult formulae"*

At the same time there were a number of positive comments such as:

- *"We need the services of a mathematician as he is someone who has a very high mental faculty and uses it to invent formulae and theorems which are later used so as to make the students develop their sense of reasoning".*

The drawings made by the children, small comments put in their drawings and reasons that they put forward to justify the services of a mathematician provide an important insight of the picture that the subject has in the mind of the student. For instance, the drawing illustrated in Fig. 5.7 on Page 160 tends to show that this student sees a person involved in doing mathematics as carrying out strange activities, with extra powers (*my brain is a computer*) and which needs a lot of effort and hard work (*I will become mad if continue to do mathematics*). It also seems that that child is describing his/her teacher, as it is shown that the teacher is giving the page numbers for homework. At the same time the teacher is uttering the words *"stupid fellow"*. This suggests that the student who has made this drawing found his/her teacher to be frequently using admonishing words. Many interpretations of these kinds can be made from the drawings made by the students.

Interview of other key informants

After having given students, teachers and parents the opportunity to express themselves on this important issue of gender and mathematics, it was considered important to obtain the views and opinions of other key informants. In fact, amongst the five people who were identified, three had many years of teaching experience at the secondary level (around 30 years) and a rich experience in teacher training at primary and secondary levels. The fourth one was a university lecturer, and the other one was lecturing at the tertiary level and also directly involved in national examinations both at primary and secondary levels. Interviews were conducted individually and their responses transcribed. A copy of each of these interviews is available in Appendix Eleven. An analysis of their responses has been made theme-wise (based on the questions) and proper quotes have been inserted in italics together with a key. For instance, **A12, I3, p.305** refers to Appendix Twelve, Informant three and found on Page 305.

You have been involved in the teaching of mathematics at various levels for years. Could you tell us about your experience as a student of mathematics and as a lecturer?
All the stakeholders interviewed had a very enjoyable involvement with mathematics since primary level. This was expected as they had all chosen the field of mathematics for their professional career and thus should have had positive attitude and performed well in the subject.

- *"I will not hesitate to say that mathematics is a very enjoyable subject, having within it its language and mode of thought, thus giving me an opportunity to appreciate the importance of precision, accuracy and rigour in our life."* (A12, I2, p.313)
- *"I found it easy to acquire the problem solving skill as the teaching method consisted in drill exercise. The teacher explained a chapter, and gave as homework a hundred problems. Only the last 30 problems were really intricate and could be solved by the bright students. I adopted the attitude that each and every problem in mathematics possess as a challenge to me. I will not rest unless I solved it."* (A12, I3, p.316)

At the same time, three of them did mention the difficult areas, difficulties they encountered while proceeding with their further studies in mathematics or the performance of students in Mauritius in general.

- *"I always enjoyed mathematics and solving problems. I never used to give up when encountering difficult problems. We used to do group work to solve problems. My favorites in mathematics were Pure Mathematics, Statistics and Numerical Analysis. However, I never used to like Mechanics."* (A12, I4, p. 320)

- *"I will talk of my experience as a student at degree level after the Higher School Certificate. I had not been exposed to the abstract nature of mathematics. In my first year at university I often wondered whether I was doing mathematics or literature. At one point I even thought of giving up. I mention this because the gap between A-level mathematics and degree level mathematics is too wide. It is getting wider now."* (A12, I5, p. 322)

- *"When I was a student in a state school, I was under the impression that the level of mathematics is very high in this country. The same thing when I was at the university as a lecturer. I was under the impression that we are among the best and all other students did well in mathematics in this country. But unfortunately as an RDO I had the hard facts, the figures in front of me; I realized that I was not actually dealing with everybody in this country when I was at the university. But when I had the figures for the whole country I found that over 55% of the students at 'O' level do not even earn a credit in mathematics. In fact around 35% fail, 20% get a pass and only 45% earn a credit, that is, get a grade from 1 to 6. Definitely there is a big problem in the teaching of mathematics in this country. There is a big problem."* (A12, I1, p. 309)

What do you believe is the trend in the performance of students in mathematics in Mauritius?

There were mixed responses to this question; some noting that there has been a slight improvement in the performance of students while others believe that there has been a decline. However, the level of achievement was commented on. Nearly all stakeholders agreed that understanding of mathematical concepts on the part of students is still a concern in Mauritius. Several issues mentioned earlier related to this: the syllabus, efforts of students, and teaching methods adopted in mathematics classrooms.

- *"I will say that at O level the percentage of students who have been successful in mathematics has been practically a constant; those getting a credit ranging from 43% to 45% and those passing from 60% to 65%. One of the reasons for this is that the syllabus has not changed for a long time. The teachers have come to know how to teach the syllabus, I suppose; very well in the sense that, though they mostly adopt the algorithmic approach, they are quite conversant with the syllabus."* (A12, I1, p. 310)

- *"I will not hesitate to say that when it comes to percentage passes they are doing quite well. But to get a very good result does not necessarily mean that they know the subject thoroughly. This trend is still there. Teachers are more concerned with getting good results but not necessarily having a deep understanding of the concepts."* (A12, I2, p. 314)

- *"The trend in the performance of students in mathematics in Mauritius is following a downward path... Students have developed an attitude to always opt for the easy path. Mathematics is not easy for those who are not working regularly and not doing sufficiently. The nature of mathematics is such that it presents itself as a chain, and if it happened that you have missed one concept, then it is difficult to understand. Mathematics is one of the subjects that use extensively the known part to develop hierarchically".* (A12, I3, p. 317)

- *"I can say that the HSC results for the intake of Year 1 at the University have decreased. Students 5 years back seemed to do much better than those of today. Nowadays, students aim at solving problems mechanically with not much of reasoning"* (A12, I4, p. 320)

- *"At Higher school level the performance is better because the syllabus has been watered down. I would tend to believe that it has also improved at SC level. This does not mean however that students are better in mathematics now than in the past."* (A12, I5, p. 322)

What do you think are the difficulties that our students encounter while learning mathematics at the secondary level?

One point mentioned very often was the algorithmic way of proceeding to deal with mathematics and the absence of conceptual understanding. The lack of opportunities for students to be involved in the thinking process has also been commented upon. Another problem that was encountered by students was language: they had difficulties in solving word problems and also with the language of mathematics itself.

- *"The very first thing is that our students have been greatly influenced by the algorithmic approach they have been exposed to and, therefore, when mathematics is not presented in the algorithmic way, they have a lot of problems in appreciating it ... Second is about mathematical language; students have a lot of difficulties in appreciating the language of mathematics. (A12, I1, p. 311)*

- *"The students have not developed the habit of reading, of looking for things on their own. They have been trained in that way, to wait for the teacher to explain everything...Also, they have problems with the language and also with the symbolic nature of mathematics. They cannot use the symbols and notations in the appropriate manner. Also, their logical argument is quite poor because they have not been preparing themselves for this". (A12, I2, p. 314)*

- *"Students do not understand the real notion of the topic; they have a procedural way of looking at the subject. Also, English language not being our mother tongue creates some problems, specially in the low performing schools. (A12, I4, p. 320)*

- *"The difficulties are understanding the basic concepts. The common mistakes are application of linearity to all situations such as ln(x+y) = lnx + lny, ln(x-y) = lnx - lny etc. Also they have difficulties in understanding topics like transformation geometry."* (A12, I5, p. 322)

Is there anything particular to boys or girls?

Almost all the stakeholders (including the woman lecturer) mentioned in some way or the other that boys have a slight edge on girls as far as involvement in mathematics is concerned. Some commented on the quality of work done by boys; the extra effort girls needed to put in to achieve in mathematics, and the gender bias in our textbooks.

- *"We can say that at primary level, girls tend to do than boys. But this trend does not continue at secondary level, unfortunately, where boys and girls do equally well. However, the quality of work is on the boys' side; that is boys provide more quality work. What I mean by that is that when it comes to the difficult parts, the boys actually do better. Though they do not excel in all domains, they certainly do really well in certain domains.* (A12, I1, p. 311)
- *"From my experience, I find that girls are more meticulous, more careful, more serious as far as mathematics is concerned. The boys take it more lightly. Also, girls are more concerned about their future."* (A12, I2, p.314)
- *"The textbooks very often are gender-biased. Most of the examples are for boys. It is worth mentioning that mathematics textbooks are mostly written by men, so we would expect more examples from boys. If we consider the Mauritian case we have more male math teacher than female (MIE all male lecturers in mathematics department) so the explanation tend to be more for male students"* (A12, I3, p.318)

- *"I feel that boys get more problems in language understanding as compared to girls. Also, girls study longer hours to achieve lesser than boys."* (A12, I4, p. 320)
- *"If latest researches are to be believed the answer is yes. According to the latest researches, brain structure of girls makes them do language better than mathematics whereas that of boys makes them do mathematics better than g. It is surprising that at primary level results indicate the contrary whereas at secondary level results favour the results of the researches. However there are cases where girls do extremely well in mathematics."* (A12, I5, p. 322)

What can you say about the mathematical problem-solving skills of boys and girls?

All the stakeholders were of the opinion that boys were better than girls as far as mathematical problem solving skills were concerned. It was also noted that boys were more adventurous — more prepared than girls to try out different methods other than those taught in the classroom, while girls tended to restrict themselves just to those methods dealt with in the classroom.

- *"As I have just said, when it comes to difficult situations, boys do better."* (A12, I1, p. 311)
- *"Boys are better situated than the girls, because they have all the prerequisites for problem solving approach. Girls want more things that are ready made."* (A12, I2, p. 315)
- *"Boys are more to the point while girls tend to be more descriptive. I can also say that the aptitudes of boys to solve problems are better than for girls. While solving problems, boys tend to try methods other than those taught in class and they come out with good results."* (A12, I4, p. 320)
- *"I think if IQ's are compared girls do as well as boys. If results of researches mentioned in above are correct then boys should be better than girls at problem solving".* (A12, I5, p. 323)

Do you feel that peers have a great influence on a student learning mathematics?

All the stakeholders were unanimous in their view of the positive contribution of peers in a student's learning of mathematics. While some shared their own experiences at school, others noted the ease at which students can communicate with each other.

- *"Definitely, my own experience at the secondary level; I was in a group where practically everything liked mathematics. I remember all of those friends did well in mathematics. In fact there were some friends in that group who were having some difficulties in mathematics. However, being with those who were doing well and with this support, they were motivated to work and ultimately all of us did well in the HSC examinations."* (A12, I1, p. 312)

- *"I do agree, I believe that they are more at ease to converse, to expose their difficulties with their friends. With the teacher they may feel embarrassed but not in their peer group."* (A12, I2, p.315)

- *"Yes I believe that peers have a great influence in a student learning of mathematics. Group work, collaborative work accelerates the process of learning".* (A12, I3, p. 318)

- *"I will say yes. We note that that good students tend to cluster together."* (A12, I4, p.321)

- *"Yes. Students who are very good may help weaker students. This is why group work can be very helpful. However when classes are very large it is difficult to do group work."* (A12, I5, p. 323)

What about the contribution of parents?

Here also, the positive contribution of parents in the teaching and learning of mathematics was noted by all the key informants. While one was talking about his own research conducted at the primary level in Mauritius, others were talking about the support parents can provide at home helping in mathematics itself. Problems that parents encounter in helping in the education of their children were also noted.

- *"Our own research at primary level shows that the contribution of parents is great. There are parents who invest in teaching aids and this definitely helps in the learning of mathematics. There is also the level of education of parents, trying to help the students in understanding a concept or a problem. This definitely helps the child. Showing an interest in the education of the child itself plays an important role."* (A12, I1, p. 312)

- *"I have a feeling that parents, in general, do not give enough time to their sons or daughters, may be for various reasons. They think everything depends on the teacher; the teacher is the sole responsible for the education of their children or may be the parents do not have enough time from their own responsibilities associated to their work. Third, because they feel they are not at ease in the modern education mode and thus do not give enough time to their children."* (A12, I2, p. 315)

- *"Parents' contribution can be considered generally as positive. They will have to provide the necessary support, moral and physical."* (A12, I3, p. 319)

- *"Home upbringing matters a lot in the study of a student. Yes, parents do have an important role in the education of their children. The monitoring of student's performance does have an effect. If parents are too strict, it has a negative impact on the child.* (A12, I4, p. 321)

- *"Parents can only help if they themselves are good in mathematics. This is true for only a few of them. Syllabuses undergo changes and very few parents can make a worthwhile contribution.* (A12, I5, p. 323)

What can be done to enhance the teaching and learning of mathematics at the secondary level?

Almost all the suggestions were related to the teaching methods to be used in the mathematics classroom. The involvement of students in learning through activities to enable development of conceptual understanding was recommended, together with group work.

- *"I believe that this algorithmic approach of teaching mathematics has been there for 30 years, we cannot expect things to change in one or two years. It will take a long period. To start with we've got to work together with the teachers. Show to them how the teaching could be done in a different way, apart from the algorithmic way, so that we can help all the children, even the bright ones, to learn mathematics. I think that this is the first thing to do; to help the teachers change the way they teach mathematics..."* (A12, I1, p. 312)

- *"Teaching in context, more trained teachers. Student-teacher ratio 15:1. Differentiation in learning, math club at school, with many mathematical activities. Make mathematics more interesting."* (A12, I3, p. 319)

- *"Initiate group working. Also, instead of tests, introduce project work as assignment. Also, teaching should be more learner - centered."* (A12, I4, p. 321)

- *"Teachers should stop teaching only to pass examinations. News from classrooms indicate that very little creative teaching goes on. The only objective seems to be able to do certain tasks dictated by the syllabus. On the other hand students are to share the blame for this situation for they only wish to learn what is in the syllabus. Projects should be included in the syllabus to encourage creativity. For this to happen, pressure must be exerted on the examination board."* (A12, I5, p. 323)

Do you feel that gender equity in mathematics education has been achieved in Mauritius? If no, what do you think can be done to achieve it?

The stakeholders agreed that gender equity had been reached to some extent in Mauritius as far as mathematics education was concerned. However, it should be noted that reference has been made mostly to equality of opportunities, while gender equity goes much beyond that, to equality of outcomes (Fennema, 2000).

- *"I feel that it has been achieved as there is equality of opportunities for the students to study."* (A12, I4, p. 321)
- *"In learning mathematics we find no difference in the nature of boys and girls – as the mind only is in play. The environment in which they develop is somewhat different. So in the teaching of mathematics this difference should be catered for..."* (A12, I3, p. 319)
- *"I tend to believe that gender equity has been reached when you look at the number of boys and girls at the HSC and BSc levels. It may not have reached an equal status but it has taken the tendency."* (A12, I2, p.316)
- *"In terms of teaching I'll say yes. There is no distinction between the teachers who teach boys and those who teach girls."* (A12, I1, p. 313)

Summary of key informants' interviews

The key informants who had many years of teaching experience both at secondary tertiary levels believed that the performance of students in mathematics was really a cause of concern. The method of teaching employed in the mathematics classrooms was mentioned to be playing an important role in this situation. Students were not given enough opportunities to develop critical thinking and to develop strategies to deal with novel situations. They had the tendency to develop procedural ways to tackle stereotyped questions without having a proper understanding of the underlying mathematical concepts. Language also was mentioned to be acting as a difficulty while solving word problems. All the key informants were of the opinion that boys have a slight edge on girls as far as involvement in mathematics is concerned. One of the most important ways to remedy the situation concerning the poor performance in mathematics was suggested to be bringing innovative changes in the method of teaching mathematics in the classrooms. Innovative ways where students are motivated and will be encouraged to be involved in critical thinking were proposed.

Answering the Research Questions 1-3

In this section the research questions posed in the beginning of the study will be answered based of the findings obtained from the two phases of the study (Chapters 4 and 5). For the first phase a survey method was used and involved 17 schools across the island in order to obtain a general idea of the performance of boys and girls across Mauritius. Data were collected mainly from two questionnaires: one specially designed for the study and the other one a modified version of the Fennema Sherman Mathematics Attitude Scale. Four secondary schools were selected (one single boys', one single girls' and two coeducational schools) for the second phase of the study. In this phase, data were collected though classroom observations, interviews with students, teachers, parents, rectors and other stakeholders in the educational sector. The two questionnaires used in the first phase also were administered to the different sample of students in the second phase. The students were also asked to fill in the Questionnaire on Teacher Interaction with a view to obtain their perceptions of the type of qualities their teachers possessed and their opinions on the type of teacher interaction present in their class. The three research questions are answered on data obtained in these two phases.

Research Question One: What are the factors that contribute to the mathematical achievement of Mauritian students at the secondary school level?

The main factors that contribute to the mathematical achievement of Mauritian students at the secondary level were found to be attitudes towards mathematics, students' mathematical background, the teaching method and influence of parents and peers.

Attitudes

In the first phase of the study a correlation coefficient of 0.336 was found between attitude towards mathematics and the total score in the mathematical test. Though rather low, this correlation suggests that there is some relationship between attitude towards mathematics and test achievement This does not mean that the better the performance in the test, the attitude towards mathematics should have been positive; however, from the interviews, it was found that students expressed their views that performance in mathematics can be improved through hard work, practice and effort. Also, the more a student likes a subject and has a positive attitude towards it, the more time he/she will devote to the subject and devote greater effort to it. It follows that, subsequently, this will allow the student to perform better in a test.

Students' mathematics background

Evidence from the survey and the interviews conducted show that another factor which impacts on the mathematical achievement of students is their own mathematical background. Mathematics is a subject which is hierarchical and sequential, and a mathematical concept has many preconcepts which normally are covered in lower classes. If a child has not properly mastered the mathematical concepts in lower classes, he/she may face difficulties to understand related topics in subsequent classes unless he/she made the necessary effort to fill the gap. In Mauritius, there is no proper syllabus with the objectives laid down for Form I, II and III. Those objectives that are covered in the topics are normally dealt with in the classes. Each school may have its own scheme of work for the year which is then broken down into three terms. It may happen that a topic which is found in the textbook is not covered at a particular level and this gap will remain until the student reaches From IV or Form V. As already mentioned, when a student has missed a topic or has not mastered a topic he/she may face difficulties in other related topics. For instance if a child has difficulties in computing $-1 + 5$ or $-4 + -3$, that student will have problems dealing with simplifications of say $-x + 5x$ or $-4y + (-3y)$ in algebra.

Teachers and Teaching method

Teachers play a fundamental role in the liking/disliking of mathematics and achievement in the subject. This study has revealed the importance of this key partner in the education field. Many students who were interviewed attributed their good performance in mathematics to their teachers. These teachers acted as role models and played an important role in encouraging and motivating students towards mathematics learning. On the other hand, there were students who attributed their failure in mathematics to their teacher. They asserted that they were discouraged in mathematics by the attitude of their teacher or the comments he/she made.

It was also found that, in general, the method of teaching adopted for mathematics at the secondary level in Mauritius is teacher-centered. The system of education in Mauritius is highly examination oriented and consequently affects the way mathematics is taught at the secondary level. The emphasis is on the completion of the syllabus and working out as many past examination papers as possible for the School Certificate and Higher School Certificate examinations. This is to the detriment of conceptual understanding on the part of students. They are rushed through the different concepts, and algorithmic approaches to solving problems are emphasized. This situation in mathematics and science education reflects that of most parts of Africa (Kogolla, Kisaka, & Waititu, 2004). Different initiatives are being taken in the African region to remedy the situation, such as the organization of in-service training sessions, collaborative research and regional conferences.

Teacher Training

One of the prominent problems related to teaching in Mauritius is that, at secondary level, there is practically no pre-service teacher training. People who complete their education can apply for the job of a teacher and start teaching in schools straight away. It is while being in their job that they join the Mauritius Institute of Education to undergo professional development involving pedagogical issues. The main criterion used for selection for the course is the number of years of teaching experience and seniority in the school the teacher is working. There are cases in which teachers have come to follow courses at the MIE after having spent some twenty years in a secondary school. Teachers tend to learn from their colleagues and develop strategies related to the teaching and leaning process. At times they develop some habits which become quite difficult to change. The need for pre-service training (which is compulsory at the primary level in Mauritius) at the secondary level is strongly felt and indeed some measures have been taken in that direction. A new course, the Post Graduate Certificate in Education, which existed only for practicing teachers, has been available for fresh graduates on a one year full time basis since 2002. The importance and relevance for such a course is yet to be established as the enrolments in this course were far from satisfactory until recently.

Parents

Parents do play a very important role in the education of their children. The students do realize this and have acknowledged it to a great extent in the interviews. Education is not just confined to the four walls of the school but takes place much beyond them. Parents have been playing an important role in the informal education of their children prior to their schooling, at the primary level and also at the secondary level. Some children claimed that the mathematics curriculum has changed so much that their parents cannot do the mathematical problems and thus cannot help them much. Parental help does not mean doing the mathematical problems for the children; it means providing the necessary moral, financial, social, psychological support.

Peers

The influence of peers was found, through the different interviews conducted, to be a factor affecting achievement of students in mathematics. The students of Form IV are adolescents and consequently peer influence is present. At times it may happen that a student has understood the explanation better from his/her friend than from the teacher. This helping attitude has been found to be amongst our secondary students. The quest amongst students to share and help does enhance the teaching and learning of mathematics. It was also found that peer influence can act in some other ways also, more among the boys — for instance, the competitive environment in the classroom. The good performance of a student in mathematics can motivate others to work harder and consequently improve other students' performance in mathematics.

Research Question Two: What types of difficulties do Mauritian boys and girls encounter while learning mathematics at the secondary level?

The previous research question was devoted to the factors that affect the learning of mathematics. The objective of this research question was to find out the difficulties the students in Mauritius encounter while studying mathematics at secondary level. It has been noted that their performance in mathematics (in terms of quality) is far from satisfactory at the School Certificate level. The importance mathematics plays in the education field, further education and the employment market in Mauritius has already been discussed in previous chapters, and the fact that a significant percentage of the Mauritian boys and girls do not reach the required standard in mathematics poses a problem for further education and job opportunities in many fields. Hence there is a need to find out what difficulties they face while studying mathematics in an attempt to remedy the situation and achieve gender equity at all levels.

Conceptual understanding

One of the foremost difficulties of many of our students is the lack of conceptual understanding of the concepts involved in mathematics. This could be noted through the survey which was carried out in the Phase One of the study. Further evidence was obtained through the interviews of teachers and the key-informants. Students tend to learn the concepts superficially to enable them to tackle the problems set on that topic. This type of teaching and learning has been going on since the primary level, and students have been used to working out past examinations papers since the age of 9 or 10 to prepare themselves for the end of primary examination. Students tend to proceed in the same way even at the secondary level, not giving themselves enough time to stop and think and make connections to previously learnt concepts. The procedural way of learning mathematics is preferred — where the students possess the necessary tools to tackle problems at the expense of proper understanding of the underlying concept. This tends to make the foundation of the mathematical knowledge weak, which further affects the building of other concepts on the original ones. This also affects them when they have to deal with problems related to real life situations, or problems involving logical thinking.

Language

Another area of difficulty relates to word problems. When the mathematics questionnaires used in Phase One of the study were analysed, it was noted that many students were encountering difficulties in solving word problems. Language seems to be a major problem while students are working with questions. This was also mentioned by students, teachers and key-informants in their interviews. The language problem seems to be more prominent among low-achieving students. English is not their first language and most of our students are exposed to at least three languages from the primary level (English, French and one oriental language) together with the one spoken by every Mauritian, the Creole. Some students do experience difficulties in analyzing a word problem, understanding what they have been given and what is to be found. It has been found that if the question is translated in Creole the students can proceed. Some examples in the mathematics curriculum where students tend to face problems are Applications of Quadratic Equations, Probability, Three Dimensional Geometry and Locus.

Research Question Three: Why do these difficulties occur?

Teaching methods

After having identified what difficulties students tend to face while studying mathematics, it was imperative to find out why these difficulties occur. One of the most important factors responsible for these problems appears to be the teaching methods used in our classrooms. From the classrooms observation carried out in this study and the interviews conducted with students, teachers and the other key informants, it was found that the method of instruction mostly used in classrooms is very much teacher-centered, with the teacher doing practically everything and the students very passive at the other end. A very algorithmic way of proceeding with mathematics is favoured, where the topic is introduced, a few examples carried by the teacher on the board, and problems set for students to try both as classwork and homework. The situation reflects that described by Welsh (1978) (cited in Romberg & Carpenter, 1986): Questions are being used in the classrooms, not to initiate discussions but just to check whether the students are 'understanding'. There is little student involvement or activities carried out.

Examination oriented system

The system of education in Mauritius is too examination oriented, and it is generally considered that the earlier one can start tackling past examination papers at the upper secondary level, the 'better' it will be for the student to gain practice in answering the questions. There are cases of some schools in Mauritius where there is no prescribed book for mathematics at the Form IV level — School Certificate past examination papers are used.

The better result one can get at the SC level implies a better school for admission to the Higher School Certificate, and this consequently increases the chance to obtain satisfactory result and obtain a scholarship for further studies. This fierce competition used to be in evidence with 10-11 year olds at the time of the end of primary level examination. Since 2002 the government has changed the admission to secondary schools to a regional basis, thus reducing to a great extent that cut-throat national competition at that young age.

In Mauritius the practice of private tuition is very significant. Many reasons have been brought forward, through the different interviews, to explain this situation including syllabus coverage, size of classes, or at times the students themselves.

Students' involvement

Students lack proper understanding of mathematical concepts and thus encounter difficulties to apply these to novel situations. Through the classroom observations and the interviews of teachers and key informants conducted, it was found that students are not actively involved in their learning of mathematics and in activities which involve critical and logical thinking. They are content with algorithms and skills to solve problems, and more emphasis is laid on product rather than process. The 'banking concept' of education is favoured; they 'bank' the necessary skills to solve problems to be 'cashed out' at the time of the examinations and not much is left over for transfer.

Students' lack of interest in mathematics and education in general was also mentioned as one factor responsible for their difficulties in mathematics. Not much effort and self-study are put in cognitive activities that might help them to develop conceptual understanding in mathematics. They rely too much on the teacher (classroom or private tuition one) and consequently do not put in the desired effort. It should be mentioned that ready-made solutions of past examination papers are available in the market and these may lull students into not trying hard enough to solve a mathematical problems.

Learning environment

The learning environment of the mathematics classrooms also influences the learning of mathematics. Many classrooms in the secondary schools in Mauritius are traditional with desks, chairs, blackboard but no extra facilities (except for a graph board in some cases) exist which may help in the teaching and learning of mathematics.

The opportunity for students to use ICT in the mathematics classroom is minimal, while at the same time software exist which could help to assist in the teaching and learning of mathematics and help students to visualize things that otherwise was difficult to see. ICT enables students to make conjectures about certain phenomena in mathematics, experiment with the software and then verify their hypotheses. This would give another dimension to the teaching and learning of mathematics which normally cannot be achieved with the traditional chalk-and-talk method. The lack of opportunities of using these facilities does influence the way mathematics is being taught and learnt.

Summary of chapter

This chapter has dealt with data obtained from the second part of the study. After a brief quantitative analysis of the responses to the mathematics and attitude questionnaires, the data from the QTI were analysed, and this was followed by further analyses of interviews of students, teachers, parents and key informants. Their responses were examined and discussed, with a brief summary provided at the end of each interview. The drawings made by students related to *Draw a Mathematician Test* were answered and the chapter ended with the first three research questions being answered. The main factors that contribute to the mathematical achievement of Mauritian students at the secondary level were found to be attitudes towards mathematics, students' mathematical background, the teaching method and influence of parents and peers. Regarding the difficulties encountered by students while they are learning mathematics, lack of conceptual understanding and the language problem were identified. The reasons for these difficulties were analysed while answering Research Question Three and were found to be the teaching method mostly used in the classroom, the examination oriented system of education, lack of active involvement of students in their own learning and the learning environment.

The next chapter discusses the third phase of the study which was devoted to the implementation and evaluation of a teaching and learning package. The fourth research question is answered in this chapter also.

CHAPTER SIX
The Implementation Phase

This chapter describes in detail the third phase of the study, namely the implementation phase. After having analysed the problems boys and girls encounter in general in mathematics at the secondary level and conducted a more in-depth study in four selected schools in the first phase of the study, the different stakeholders involved were interviewed for the second phase. The point of view of these stakeholders provided more information on the reasons for Mauritian students encountering difficulties in mathematics. Different factors were identified and strategies were formulated to help in enhancing the teaching and learning of mathematics at the secondary level in Mauritius. With a view to testing the efficiency of these recommendations, a number of the strategies were tested in three selected schools (one single boys', one single girls' and one co-educational) in the third phase of the study to trial the preparation of a teaching and learning package.

In this chapter the strategies adopted for the implementation phase are described together with the description of how the implementation of the package was carried out. Two lessons are then described in detail. The results of the pre-test and post-test are also analysed together with the results obtained from the What Is Happening In this Class? (WIHIC) questionnaire. Students were also asked to write a small anonymous report on their frank opinion of the strategies that were used over the three months duration of the study, and of their efficiency on the teaching and learning of mathematics. The chapter concludes with an answer to the fourth research question.

Through the interviews of students, it was found that teachers have a great influence on their attitudes towards mathematics and their achievement in the subject. It was also found that the teaching of mathematics at the secondary level in Mauritius was teacher-centered, examinations-based, and that students were passive in the classrooms. More emphasis was laid on procedural understanding of mathematics, and it was found that students were experiencing difficulties when the context of problems was unfamiliar. The same situation exists in many parts of Africa concerning mathematics and science (Afrassa, 2002; Githua & Mwangi, 2003). In a training programme organized in Kenya in November-December 2004, fifteen countries across Africa discussed the problems related to the teaching and learning of mathematics and science and the urgent need to bring about innovative changes in the method of teaching was underlined. The dry, traditional way of imparting knowledge to students was found to be ineffective in developing critical thinking and problem solving skills in students (Kogolla, Kisaka, & Waititu, 2004).

The strategies that were used in this phase of the current study to help students in their learning of mathematics were: the use of cooperative learning, the involvement of students in their own learning, and the use of activity-based teaching. The lessons that were developed which catered for these strategies were based on a philosophy developed in the Strengthening of Mathematics and Science in Secondary Education project (SMASSE) with the collaboration of the Japan International Cooperation Agency (JICA). The philosophy is explained later in this chapter.

A brief report on cooperative learning and research related to its use together with a description of the approach used for developing the lesson plans are provided below.

Cooperative learning in mathematics

As noted by Johnson & Johnson (1990, p. 104) the method used in the classroom for the teaching of mathematics plays an important role on the efficiency of the lesson:

> Having students work cooperatively, competitively or individualistically has important implications for the success of math instruction. In a cooperative learning situation, students' goal achievements are positively correlated; students perceive that they can reach their learning goals if and only if the other students in the learning group also reach their goals.

Much research has been conducted related to the use of cooperative learning in mathematics (Davidson, 1996, 1990; Khalid, 2004; Olson, 2002; Peterson, 1988; Posamentier & Stepelman, 2002). The efficiency of cooperative learning in enhancing teaching and learning of mathematics was commented upon by Davidson (1996).

> Small-group cooperative learning can be used to foster effective mathematical communication, problem-solving, logical reasoning, and the making of mathematical connections... Davidson (1996, p.52).

Furthermore the positive effects cooperative learning have on academic achievement, self-confidence of the learner and use of social skills have been noted by Davidson (1996).

Just by placing students in groups and asking them to work together does not in itself ensure cooperative learning amongst them. The teacher plays an important role in the successful development and sustainability of group dynamics. In the sessions where cooperative learning was used, I ascertained that the students within each group realized that they were part of a team and they all had a common goal. They were made to realize that their success depended on the input of each student in that group and there was a need for positive interdependence. They were encouraged to listen to other's ideas, discuss, offer and accept constructive comments and develop positive social skills. My role in monitoring and intervening also was important. While I was moving from group to group giving assistance and encouragement, I refrained from providing ready-made solutions to any call from a student. In fact, by asking appropriate questions, the group was encouraged to rethink on the strategy/method they were using and consider alternative ways of proceeding. In this way students were encouraged to engage in metacognitive reflections related to the strategies that were used for solving a particular problem. I also ascertained that each student in a group was participating fully in the discussions and that there was no dominance by or of any student in a group. The findings of each group were displayed on the board on a Bristol paper, and I ascertained that there was agreement and rotation concerning the student to present the work in front of the class.

ASEI movement & PDSI approach

In its main aim of enhancing the teaching and learning of mathematics and science, the ASEI (Activity, Student, Experiment, and Improvise) Movement initiated the Strengthening of Mathematics and Science in Secondary Education project (SMASSE, Kenya) with the following underlying principles:

(a) Knowledge-based teaching to be replaced by activity-based teaching,

(b) Student-centered learning to prevail over teacher-centered teaching,

(c) Experiment and research based approaches to replace the traditional lecture approach,

(d) Improvisation in experiments to eliminate the necessity to resort to large scale "recipe" type experiments.

This approach was mentioned in a UNESCO document entitled *Connect* in 2003. The meaning of each letter in the acronym ASEI is explained below.

Activity focused teaching/learning

One of the main pillars of activity-based teaching is that active learner involvement in the teaching/learning process enhances understanding and promotes retention. We have the famous proverb:

> *I hear, and I forget;*
> *I see, and I remember;*
> *I do, and I understand.*

Teachers need to shift from practices where the learner is a passive recipient of knowledge to where they participate in generating the knowledge and appropriate skills and attitudes. Appropriate activities should be identified in line with the lesson objectives and which help in enhancing the teaching and learning process (and not activity for activities' sake). Proper bridging from the activities to the underlying concept needs to be ascertained.

Student-centered teaching/learning

This aspect of the ASEI calls for a shift of classroom focus from the teacher as the main actor to the learner. The learners should be actively involved in the teaching/learning process and appropriate opportunities should be provided to them to express opinions and explain ideas based on their prior experiences and verify these through suitably designed teaching/learning activities. Teachers should act more as a facilitator, assisting learners in the construction of their knowledge. More emphasis is put in a process where students are involved in learning activities. In so doing, students will be able to construct cognitive skills which are applicable in obtaining other required skills and knowledge.

Experiments

The emphasis here is a shift from recipe type experiments to investigative types that allows learners to make predictions/hypothesis, verify them practically and where possible design their own experiments.

Improvisation

This aspect of the ASEI calls for the innovativeness and creativity of the teacher. It involves thinking of other ways of proceeding and/or utilizing locally available materials in the immediate environment of the students to raise interest and curiosity. Mathematics has been described many times as being dry and dull and very far from the real life situations. The teacher should think of ways and means of reaching the learners and making the learning relevant and enjoyable.

For the principles of ASEI to be put into effective use, the PDSI (Plan, Do, See, Improve) approach should be used. This approach basically calls for:

Plan - Proper planning of the lesson based on the ASEI principles is important. While designing the lessons for the implementation phase, the learners' background, needs, interests, misconceptions and prior knowledge related to the lesson content were taken into consideration. Instructional activities were then identified to help learners to empower their existing knowledge by relating the new knowledge to existing concepts and forming appropriate connections. This knowledge was meant to be used for transfer purposes also — that is, retain the learning and apply it in other contexts and real life situations.

Do - This aspect is basically concerned with lesson delivery; the instructional process based on the plan. An appropriate lesson introduction was designed for each lesson to the students' interest. The introduction activity allowed relating to the learners' previous experience and provided an orientation to the lesson objectives. It also helped me to quickly establish what the students already knew about the lesson content, identify and deal with any misconceptions on the part of the student to ascertain the links to be formed amongst the existing and new concepts. The link of the lesson content to real life experience at this stage helped in arousing and maintaining the interest of students during the lesson. They were then encouraged to work on the identified activities in groups, which were sufficiently varied and interesting to motivate the learner's engagement and to facilitate meaningful learning experiences. These activities were meant to facilitate growth of process skills such as observing, measuring, identifying variables and planning experiments. As described earlier, I dealt with students' questions, misconceptions and reinforced learning at each step to develop critical thinking. By asking probing questions, I also helped in deriving proper meaning out of the lesson activities for students and ensured proper bridging between the activity and the underlying concept.

See - This aspect deals mainly with the evaluation both as the presentation progresses and at the end using various techniques and feedback from students. Evaluation was an integral part of each of the lessons and this was carried out to ascertain that the objectives of the lesson were met and meaningful learning was taking place. In case of difficulties, proper intervention was made to assist the student or group in their task. Assessment opportunities to monitor student progress during and after lesson were provided.

Improve - On the basis of feedback obtained in the See component of this approach, appropriate actions were brought in during the development of the lesson and/or in subsequent lessons. After each lesson, I reflected on the information obtained and brought about changes to further enhance the teaching and learning process.

The sample

Three schools (one single boys', one single girls' and one coeducational) were selected for this phase of the study in such a way that they all were comparable in terms of percentage pass at the School Certificate level and the reputation they had in the educational field. Letters explaining the objectives of the research and the purpose of this intervention programme were sent to each rector of the schools selected and an appointment made. During the meeting with the rector, further details related to the project were provided, and once the approval of the rector was obtained, the Heads of the mathematics departments in the school were contacted. The objectives of the research were explained to them by the rectors and each one was asked to choose a Form IV class. A meeting with the classroom teacher was held to explain the purpose of this study and the same explanation was given to the students in each of the classes.

A preview of the lessons

After having explained the objectives of the study to the students, a pre-test was administered to the students in each class to gain an insight of the extent to which the students in the sample had an understanding of the concepts dealt with in the lower secondary classes. The pre-test consisted of eleven questions based on the fundamental concepts which are prerequisites for further development of mathematical concepts dealt with at Form IV and V levels. A copy of the pre-test is found in Appendix Fourteen. The scripts for each of the schools were then corrected by me. After the administration of the pre-test, an intervention programme for a period of three months was commenced where the topics Quadratic Equations, Applications of Quadratic Equations and Angle properties of Circles were taught in each of the classes. The strategies that were used in the classes were: student-centred teaching, use of relevant activities, and cooperative learning. The plans that were devised for these lessons were based on the ASEI- PDSI principle. As I consider it important for the reader to gain an impression of the nine different lessons (repeated two more times in the other two schools) that I taught during the intervention period and the type of work the students were engaged in during the lessons, a running commentary on the conduct of two separate lessons is presented here in detail. The first one deals with factorization of quadratic expressions where the coefficient of x^2 is one, and the second deals with angle property in a cyclic quadrilateral. The activities that were designed to involve students' participation through group work are also discussed. Lesson plans for the other sessions conducted in the classes are given in Appendix Fifteen.

Lesson One (Factorisation of quadratic expressions)

The instructional objectives of the lesson were to enable students to:

(1) recognise a quadratic expression

(2) factorise a quadratic expression when the coefficient of $x^2 = 1$

with the knowledge of linear expressions and multiplication of two linear expressions needed as prerequisites.

As a starter, I asked students to give me an example of an expression. Students responses were written on the board followed by discussions.

Students were made to realise the importance of a variable.

Students were then encouraged to consider the power to which the variable in each expression is raised. The concept of degree was then introduced.

The following expressions were then written on the board:

$x^2 + 3x + 2$ $\qquad\qquad$ $2m^2 - m + 1$

$4x^2 - x + 1$ $\qquad\qquad$ $4 - n - 3n^2$

$y^2 - y + 15$

Through discussions, students were made to understand that the degree of each one was two; and the term **quadratic expression** was introduced.

The general form of a quadratic expression was stated as $ax^2 + bx + c$ where $a \neq 0$. (Through examples students were convinced of the importance of the condition $a \neq 0$).

Students were then asked to give more examples to quadratic expressions.

Student activity 1: Students, grouped in batches of five or six, were asked to expand $(x + 3)(x + 4)$ [These have been covered in lower classes].

I walked around to ensure that the answer $x^2 + 7x + 12$ was obtained in each group. In case of any difficulty within a group, I provided assistance by discussing with the students.

I then stated that the objective of the present lesson was the reverse operation of the expansion process, called *factorization*.

Discussion with students what it meant to factorise the expression $x^2 + 5x + 6$ was then initiated.

Student Activity 2: Each group of students were provided with the following cut-outs from Bristol paper, the first one representing x^2, each of the strip representing x and each small square representing a unit.

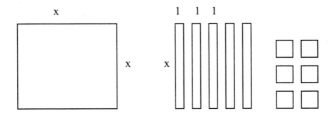

A picture of one such set is shown below in Figure. 6.1.

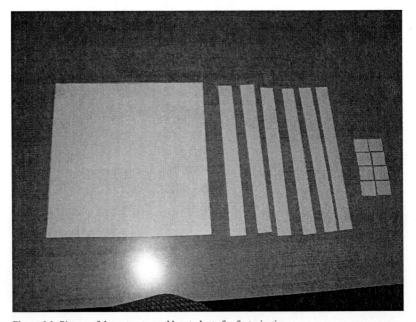

Figure 6.1: Picture of the cut outs used by students for factorisation

As a group, students were asked to manipulate the cut outs provided and to make a rectangle by using **ALL** the given pieces.

Some possibilities were:

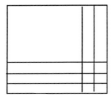 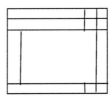

Students were then asked to find the area of the rectangle they have constructed in terms of x and to note their answer.

The findings of the different groups were displayed in front of the class together with the area of the rectangle.

Through discussions, and help, students were encouraged to compare this area with the total areas of the cut outs provided and to deduce that

$$x^2 + 5x + 6 = (x + 2)(x + 3)$$

As another activity, students were asked to work in groups and to factorise $x^2 + 8x + 15$.

Students were then asked to discuss in their groups and find the relationship, in each case, connecting coefficient of x, constant term and constants in the factorised form.

1. $x^2 + 5x + 6$ $\qquad\qquad$ $2 \times 3 = 6$

$\qquad\qquad\qquad\qquad\qquad\quad$ $2 + 3 = 5$

2. $x^2 + 8x + 15$ $\qquad\qquad$ $3 \times 5 = 15$

$\qquad\qquad\qquad\qquad\qquad\quad$ $3 + 5 = 8$

Discussions on these relationships then followed.

They were then helped to find out that

Product of the two numbers = constant term in the quadratic expression and

Sum of the two numbers = the coefficient of x.

Students were then asked to factorise $x^2 + 9x + 20$.

Two numbers that multiply to give 20 and add up to give 9 are 4 and 5.

Hence, $x^2 + 9x + 20 = (x + 4)(x + 5)$.
Also, $x^2 + 4x - 12 = (x + 6)(x - 2)$.

After making a summary of the main points of the lesson, with the help of students, and providing more clarification where needed, the following expressions were given to be factorized as evaluation:

1. $x^2 + 9x + 18$ 2. $x^2 + x - 72$ 3. $x^2 + 12x + 27$
4. $x^2 - 2x - 35$ 5. $x^2 - x - 72$

Comments on the class

The students appeared to be excited with the idea of working in groups. They normally did not work like that and the furniture had to be moved to enable cooperative learning to take place. The students took some time before they settled and were ready for the lesson. The activity where they had to manipulate Bristol paper cutouts to help them in factorizing some quadratic expressions proved to be exciting for the students (and the classroom teachers too). They then quickly reverted to the usual way of finding the factors as an introduction to quadratic expressions had already been covered in Form III. There were meaningful discussions within the group and in the whole class also. It should be mentioned that in the coeducational class, very few of mixed groups (boys and girls together) were formed. Upon my request, some students changed places and worked collaboratively with the other students in the group. While conducting the evaluation throughout the lesson and particularly towards the end of it, I found out (through observation and questions) that the objectives that were set for the lesson were met. It was also noted that the students enjoyed their own involvement in their learning. An overall appreciation of the lessons can be noted through the self-report submitted by the students, a sample of which is given in Appendix Nineteen.

Lesson Two (Angle properties in a cyclic quadrilateral)

The instructional objectives of the lesson were to enable students to:

(i)   identify a cyclic quadrilateral

(ii)   deduce (find out) that opposite angles in a cyclic quadrilateral are supplementary

(iii) apply the above result to solve problems.

The prerequisite knowledge for this session were:

- Angle subtended by an arc (chord) at the circumference

- Angle in a straight line = 180°

As a starter, questions related to the previous lesson on the relationship between the angle subtended by an arc (or chord) subtended at the centre and that at the circumference were asked of the pupils. A brief overview of the present lesson's objectives were then made.

Student activity 1:

Students, in groups, were asked to draw a circle on a bristol paper and choose four points A, B, C and D as shown (teacher showed the points, approximately on the board).

They were then asked to join A to B, B to C, C to D and D to A

Teacher: What shape is drawn?

Expected answer (on an individual basis): A quadrilateral

Question to students: Where are the vertices of the quadrilateral?

Expected answer: On the circumference of the circle.

I then introduced the concept of cyclic quadrilateral.

A cyclic quadrilateral is a quadrilateral where all the vertices lie on the circumference of a circle.

Student Activity 2:

Students, in group, were asked to measure angle ABC and angle ADC and note their answers. What do they notice?

The results of each group were to be displayed in a table.

Group	Angle ABC	Angle ADC

Students were then asked to deduce what they noticed from the table.

Student Activity 3:

In each group the students were asked to label the angle ABC as x and the angle ADC as y, as shown in Figure. 6.2.

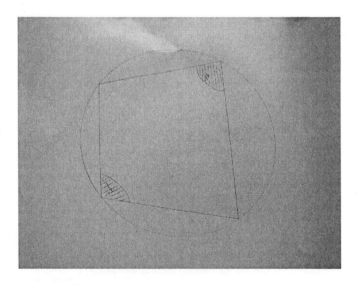

Figure. 6.2: A picture of a cyclic quadrilateral drawn with the opposite angles shaded

Using a blade, students were asked to cut out the angles ABC and ABC (in each group).

By drawing a straight line and placing the cut-outs with blue tack, students were encouraged to note what they observed.

Through discussions, students were encouraged to note that once they had placed the cut out of angle ABC on the line, the other cut-out fitted exactly.

Figure. 6.3: The cut outs fit exactly on a straight line

Question to students: What can you deduce?

Students discussed within their groups to find out that
$$x + y = 180°$$

Questions to students: What can be said about the sum of angles DAB and DCB?

Through discussions, among their groups and in whole class discussions, students were encouraged to deduce that the sum of opposite angles in a cyclic quadrilateral is 180°.

Student Activity 4:

Students were asked to draw a circle and a cyclic quadrilateral on a bristol paper as before.

Students were encouraged to discuss within their group to establish the above result. (Hint: use angle at the centre)

$ABC = x^\circ$

$AOC = 2x^\circ$

Reflex angle $AOC = 360^\circ - 2x^\circ$

But $ADC = \frac{1}{2}$ reflex angle AOC

$y^\circ = \frac{1}{2}(360^\circ - 2x^\circ)$

$\quad = 180^\circ - x^\circ$

Hence $x^\circ + y^\circ = 180^\circ$

B

A

. O

C

D

After making a summary of the main things developed in the lesson, the following problems were set as evaluation:

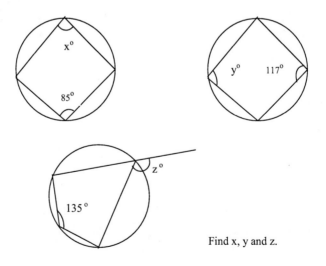

Find x, y and z.

220

Comments about the class

By that time students were quite at ease with the idea of cooperative learning and doing activities in mathematics classes. Chair arrangements were usually decided before the start of the class, and students had all the materials necessary ready for the class because they were notified in the previous session of the materials to bring to the class. It was also noted that at times there were students who were giving the property related to cyclic quadrilaterals prior to the activities. This was so as they had already covered that topic in private tuition. They knew the property but they did not why it was so. In any case, they were interested to discover certain things that they did not know. In short, I can say that the students were seen to be enjoying the learning of mathematics, and they could also attempt successfully the questions which were set for evaluation. It was also noted that some students had difficulties in handling their mathematical instruments and had not yet developed their manipulative skills. They were very often helped by their friends in their group, and in some cases by me.

Post-test

At the end of the implementation phase, a post-test consisting of fourteen items in all were administered to the students in each of the three schools. The first eleven questions were similar to those of the pre-test and these were used for comparison purposes between the pre-test and the post-test. The other three questions were based on topics that were taught during the implementation phase namely Quadratic Equations, Applications of Quadratic equations and Angle properties of circles. The performances on these three questions are assessed separately. The results of the pre-test and post-test for the three schools are given in Table 6.1.

Table 6.1: Means and standard deviations for pre-test and post-test

	Pre-test		Post-test	
	Mean (out of 50)	Standard deviation	Mean (out of 50)	Standard deviation
School 3A	8.5	6.33	17.4	5.80
School 3B	18.8	8.91	27.3	5.79
School 3C	30.3	10.10	35.2	6.48

Note that in each of the three schools there has been considerable improvement in the performances from the pre-test and the post-test. However, it can be also noted that the initial level of the three classes were not comparable: the mean performance of the students in School 3C (a single girls' school) in the pre-test was much higher than that of the students from the School 3A (a single boys' school). It seems that the students of School 3C were performing better in mathematics than the students of the other two schools. The performances of the students in the three schools in both the pre-test and post-test are presented graphically in Figure. 6.4 below.

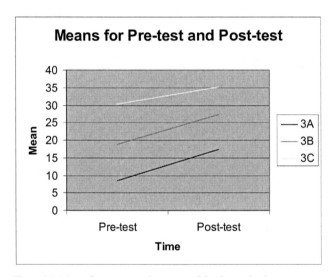

Figure 6.4: Means for pre-test and post-test of the three schools.

To find out whether the differences in the two tests were significantly different, the Wilcoxon Signed Ranks Test were performed (as the marks were not normally distributed) and the results displayed in Table 6.2 below.

Table 6.2: Results from the Wilcoxon Signed Ranks Tests

School	z value	Significance
3A	-4.646	0.000
3B	-4.796	0.000
3C	-4.176	0.000

It can be noted that in all the three cases the differences were significantly different. This suggests that the intervention programme had a positive influence on the mathematics achievement of the students as far as the basic topics in the lower secondary mathematics curriculum were concerned.

As mentioned earlier, the second part of the post-test was devoted to questions on the topics which were dealt with in the implementation phase, that is, Quadratic Equations, Applications of Quadratic Equations and Angle properties of Circles. The results of the students from the three schools are given below in Table 6.3.

Table 6.3: Means and standard deviations in the second part of the post-test

School	Mean (out of 20)	Standard deviation
3A	8.03	3.81
3B	10.11	3.94
3C	16.1	2.96

Here also the difference in performance of the students in the different schools can be noted as was the case in the pre-test and the other part of post-test. Students of School 3C (single girls' school) performed well even in this part of the post-test, indicating perhaps the seriousness of these girls to this test and dedication to their studies. While the students from the other two schools showed enthusiasm and commitment throughout the implementation stage, their performance in the second part of the post-test (and in the first part also) was not up to expectation. The students of Form 3B claimed that the time allotted for the test was insufficient: they seemed to have mismanaged their time. It also seems that that since the post-test was almost towards the end of the term, the students of School 3A and 3B did not give too much attention to it as they knew that the marks were not to be taken into consideration for the first term results. They seemed to have paid more attention to the tests which were administered by the class teachers for the first term results. However, it should be pointed out that there were students in the two schools who scored 19 or 20 out of 20 in the second part of the post-test.

What Is Happening In this Class? (WIHIC) questionnaire

Another questionnaire designed to evaluate the learning environment in classrooms is the WIHIC. It contains seven scales, each comprising of ten items. The scales are: Student Cohesiveness, Teacher Support, Involvement, Investigation, Task Orientation, Cooperation and Equity. A five-point Likert response scale ranging from Almost Never to Almost Always is used for each item. A brief description of each scale together with a sample item is given in the table below.

Table 6.4: Scales and sample items of the WIHIC questionnaire

Scale name	Description of scale	Sample item
Student Cohesiveness	Extent to which students know, help and are supportive of one another	I make friendship among students in this classroom
Teacher Support	Extent to which the teacher helps, befriends, trusts, and is interested in students	The teacher takes a personal interest in me
Involvement	Extent to which students have attentive interest, participate in discussions, perform additional work and enjoy the class	I discuss ideas in class
Investigation	Extent to which there is emphasis on the skills and their use in problem solving and investigation	I am asked to think about the evidence for statements
Task Orientation	Extent to which it is important to complete activities planned and to stay on the subject matter	Getting a certain amount of work done is important
Cooperation	Extent to which students cooperate rather than compete with one another on learning tasks	I cooperate with other students when doing assignment work
Equity	Extent to which students are treated equally by the teacher	The teacher gives as much attention to my questions as to other students' questions

(Koul, 2003, p. 30)

A copy of the questionnaire appears in Appendix Seventeen.

The reliability and validity of this questionnaire has been established in Australia and other countries and have been reported in many studies (Aldridge, Fraser, & Huang, 1999; Chionh & Fraser, 1998; Koul, 2003; Rickards, den Brok, Bull, & Fisher,

2003). The WIHIC questionnaire was used in a study conducted by Rawnsley (1997) to investigate associations between learning environments in mathematics classrooms and students' attitudes towards that subject in Australia. It was found that students' attitude towards mathematics was more positive whenever the teacher was perceived as very supportive and equitable, and when the students were involved in investigative tasks. To investigate the perceptions of the students involved in this phase of the study about the different issues like teacher support, cooperation and equity, the WIHIC was administered to them and the results for each of the schools are shown below.

Table 6.5: WIHIC results for School 3A in Phase Three

Scale	Mean	Standard deviation
Student cohesiveness	4.19	0.35
Teacher support	3.30	0.41
Involvement	3.41	0.23
Investigation	3.33	0.18
Task Orientation	4.03	0.16
Cooperation	3.67	0.23
Equity	3.80	0.30

Note that the mean on each of the scales of the WIHIC questionnaire is reasonably high in School 3A.

Table 6.6: WIHIC results for School 3B in Phase Three

Scale	Mean	Standard deviation
Student cohesiveness	3.97	0.37
Teacher support	3.02	0.38
Involvement	3.10	0.07
Investigation	3.08	0.29
Task Orientation	3.98	0.27
Cooperation	3.58	0.25
Equity	3.60	0.11

The same remarks can be said for School 3B also.

Table 6.7: WIHIC results for School 3C in Phase Three

Scale	Mean	Standard deviation
Student cohesiveness	4.19	0.42
Teacher support	3.11	0.49
Involvement	3.23	0.19
Investigation	3.32	0.24
Task Orientation	4.48	0.20
Cooperation	4.08	0.24
Equity	4.04	0.22

These results were then combined to have an idea how the three schools compare in each of the scales. Figure 6.5 shows a line graph where the means in each of the scales have been plotted for all the three schools.

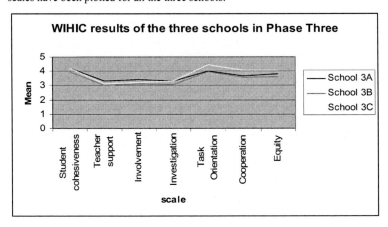

Figure 6.5: WIHIC results for schools 3A, 3B and 3C

It can be observed that the perception of the students in all the scales are more or less the same, with the students from School 3C (the single girls' school) rating higher than the students from the other two schools on Task Orientation, Cooperation and Equity. To determine the overall perception of all the students involved in this phase of the study, the data from the three classes were combined and an overall analysis made on each of the WIHIC scales. The Cronbach coefficients for each scale were also calculated to find out the reliability of the WIHIC questionnaire. The results of this analysis are shown in Table 6.8.

Table 6.8: WIHIC results for the three schools combined

	Mean	Standard deviation	Cronbach coefficient
Student cohesiveness	4.02	0.39	0.76
Teacher support	3.13	0.47	0.83
Involvement	3.22	0.14	0.82
Investigation	3.22	0.26	0.87
Task Orientation	4.18	0.29	0.74
Cooperation	3.63	0.24	0.82
Equity	3.77	0.16	0.83

One can note that the overall perception of the classroom environment was positive and the Cronbach alpha reliability coefficients were reasonably high (ranging from 0.74 to 0.87). These coefficients are comparable to those obtained in studies conducted in Australia, Taiwan, Brunei, Singapore for all the scales but are larger than those obtained in India (except for the scale Equity where they are equal) (Koul, 2003). This provides evidence that the WIHIC can be used in other studies in Mauritius with confidence.

Feedback from students

To determine the perceptions of the students concerning the strategies that were used in the teaching and learning of mathematics during the three months of the implementation stage, they were asked to write a small report anonymously. From the reports it was found that the strategy of cooperative learning was popular with the students. Some of the comments are given below:

- *"...I really liked the maths class on Thursdays because these classes were different from our usual class, that is usually in class we work personally by ourselves and for ourselves but in the maths class we worked in groups and discussed problems."* (A19, S1, p. 359)
- *"This type of teaching is actually needed in most schools as the school teacher himself does not have enough special attention for all students."* (A19, S2, p. 360)
- *"With the group work we can share our point of view with our friends on a topic".* (A19, S3, p.361)
- *"Group work means teamwork thus if I do not know a work, I can ask my friends. In this way I can share my views and opinions with them and at the same time I can learn new things".*(A19, S9, p.367)
- *"Doing work and solving problems in groups was very amusing"* (A19, S11, p.369)

- *"Mathematics classes became interesting for those who thought that maths was a boring subject, due to group works and discussions to find a problem's solution".* (A19, S7, p. 365)
- *"I never knew that we can do activities to learn mathematics properly. I never had any opportunities to do class activity in forms I, II, III".* (A19, S9, p. 367)
- *"Doing those activities were very knowledgeable as well as amusing as it was the only class in which we were doing group work and were given special attention. It was almost the only class in which we were all participating".* (A19, S2, p. 360)
- *"The class and the teaching was interesting as I never knew that we could do practical in MATHS too!"* (A19, S10, p.368)
- *"The activities in the class made mathematics more interesting to learn"* (A19, S11, p. 369)
- *"By doing activities in the class among friends was enjoyable and made things/explanations more easy to be understand".* (A19, S4, p. 362)

One student asserted that not all those in her group had the same chance of performing the activities successfully. She further added that she does not like group work as she likes to work by herself, finds her own mistake and make the corrections.

- *"It was not interesting because not all the pupils in the group had an equal chance of carrying the activities…I did not like the group work as I like to learn by myself and I correct myself".* (A19, S5, p. 363)

One of the classroom teachers attended all the sessions that I conducted in his classroom and acted as a member check in certain cases. His comments are given below:

- *"The method of cooperative learning was a new way of teaching for the students and this made them quite interested and excited on hearing that they would be working groups. As for the class, I found that the students were more willing to participate and the fact that they were working in groups, there were more communication with other friends concerning the chapter they were doing"* (A21, T, p. 370)

However, the teacher felt that the use of cooperative learning in the mathematics class may be time consuming and might delay the completion of the syllabus.

- *"I think that this way of teaching has really helped the students in one way or another. However, if this method was to be used everyday, it would take a lot of time to arrange the tables and chairs because not all classes use the same method and this can cause a problem to finish the scheme of work."* (A21, T, p. 370)

This is, in fact, the general concern of the teachers in Mauritius — completion of the syllabus. Teachers should be sensitized that this method of cooperative learning can be used successfully to fit in their time schedule and to complete their syllabus on time with students having developed understanding of the mathematical concepts.

Answering Research Question Four

Research Question Four will now be answered based on findings obtained from Phase Three of the study. Strategies that were identified after having collected data from phases one and two of the study were implemented in three secondary schools (one single boys', one single girls' and one coeducational) for a period of three months. Pre-tests and post-tests were also used to find the effectiveness of the strategies used.

Research Question Four: How effective is a teaching learning package, based on the findings of Research Question One, at enhancing the attitudes and mathematical achievement of secondary school students?

After Phase Two, where the voices of the students, teachers, parents and key informants were heard, one of the most important things that was noted was an urgent need for a different method of teaching mathematics in our classrooms. A call for more a student-centered approach, where students are involved in their construction of mathematics knowledge, was made. The students wished to take part in and responsibility of their own learning, but the present way mathematics is normally taught in secondary schools does not offer much opportunity for this to occur. A teaching and learning package was designed to cater for such an environment. Lessons were devised with a student-centered approach in mind where activities had an important role to play in the teaching and learning tasks based on the principles of the ASEI/PDSI approach. Cooperative learning was also used where students would discuss the different tasks involved in a group activity and cooperatively think of strategies to successfully perform the required task.

The role of the teacher changes from a provider of knowledge to someone who assisted students in constructing the knowledge. The role of bridging should be highlighted as at this stage proper connections are to be made from the outcomes of the activities to the underlying mathematical concept or principle. At this stage also, the role of the teacher is important: how to get students to make the connections. At the start of the intervention programme, the students took some time to adapt to the idea of activity based teaching in mathematics. They were used to activities in lower primary classes (average age 6 to 8 years) but then the activities became rarer in the upper primary classes to become almost non-existant at secondary level. The students were being exposed to the idea of performing activities in a mathematics class almost after six years. There were comments made by students at end of the intervention programme that they never knew that there could be 'practicals' in mathematics classes. But after one or two sessions, they were quite at ease with the idea and doing well on the tasks. The same can be said for the use of cooperative learning in the classes. Students took some time to understand how to go about it, but with my help they could work happily on tasks which were set within their groups. The strategies proved to be useful in motivating students to learn in mathematics by emphasising the enjoyment and confidence the students were experiencing in their mathematics classes. Students' reports concerning their feeling about the mathematics sessions also provide more evidence. The same is confirmed by the results obtained for the WIHIC questionnaires, with a high level of reliability comparable to those obtained in other international studies.

The improvement in the mathematics tests also provides evidence that the intervention programme was successful in assisting students to enhance their understanding of the mathematical concepts which were involved. Students were in fact expressing the wish that their regular mathematics classes be conducted in this way. I believe that if mathematics classes are conducted in this way where students are the focus of the classroom transactions, are involved collaboratively in performing activities and solving problems and constructing their knowledge, the attitude of students towards mathematics will be improved and their mathematics achievement enhanced. This sequence is shown pictorially in the following model.

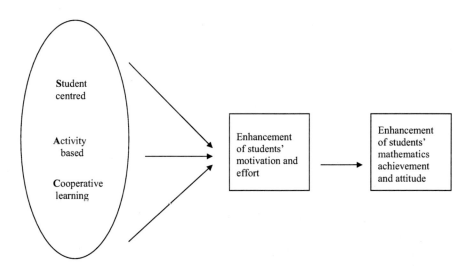

Figure 6.6: The SAC model for enhancing mathematics achievement

Summary of the chapter

This chapter has described the third phase of the study where a package was designed to help in the teaching and learning of mathematics. The package was implemented for a period of three months in three secondary schools. The framework for designing the lesson plans and the strategies used inside the classroom were also discussed here. The efficacy of the package was tested using a pre-test and post-test technique together with the WIHIC questionnaire. Feedback from students on the strategies used and the overall environment of the class were obtained through reports which were also discussed in this chapter. Research Question Four of this study was answered in the latter part of the chapter which indicated that there was an improvement in students' mathematical achievement in the post-test and their involvement in the construction of their own knowledge.

An overall discussion of the study with research findings, recommendations and prospects for future research are discussed in the final chapter.

CHAPTER SEVEN
Discussion and Conclusion

This chapter builds upon the answers to the research questions which were reported in Chapters 5 and 6. The first section of the chapter provides a brief review of the study, then the main findings and their significance are discussed. The implications of the findings for the teaching and learning of mathematics to improve the achievement of girls is then addressed. Because any educational research is normally carried out in a particular setting, researchers do have constraints related to their studies to deal with. Thus the limitations of the present study are outlined before a list of recommendations for future research brings the study to its conclusion.

Review of the study

Being involved as a secondary school teacher for eight years as well as being in close contact with students (boys and girls) in the teaching of mathematics for nineteen years in all, it was clear to me that students in Mauritius were encountering difficulties in learning mathematics. More evidence supporting this observation was obtained as I interacted with teachers across Mauritius when I became involved in presenting teacher-training courses — teachers shared their concern related to the differential performance of boys and girls in mathematics with me. It was generally felt by teachers that girls were attentive, showed interests in studies, took pride in the presentation of their work and, in general, tended to adopt strategies demonstrated by the teacher in the mathematics classroom and avoided the possible risks of using alternative strategies while solving problems. Boys, on the other hand, were described as being more active, prone to be distracted from studies and irregular with their homework. However, while solving mathematical problems, boys were prepared to experiment with different strategies to those demonstrated in the classroom.

My interest in this area, that is the differential performance of boys and girls in mathematics, grew when I analysed the results of the School Certificate examinations. I noted that boys were consistently performing better than girls at that level. While conducting a literature search, I realized that it was an area of major concern and was the subject of much research across the world.

Taking into consideration the importance mathematics plays in the Mauritian society for further studies and employment prospects, I became concerned that this differential performance of boys and girls in mathematics could be acting as a 'critical filter' in the social and economic mobility in Mauritius. With a conviction that secondary school students in Mauritius could be helped to enhance their mathematics achievement, I decided to conduct this study to identify the factors impacting on the mathematics achievement of boys and girls at secondary level. The methodology used was a mixed procedure, involving both quantitative and qualitative methods. The study, which lasted for two years (March 2003 to April 2005), was conducted in three phases: the first used a survey approach in which 607 Form IV students from seventeen schools across Mauritius (including Rodrigues) were involved. A mathematics questionnaire with items based on lower secondary concepts was administered to these students. Another questionnaire to find out their attitude towards mathematics was also used in this phase of the study. The SPPS package was utilised to analyse the responses of the students to these two questionnaires.

The second phase of the study introduced a case study method where four schools (one single boys', one single girls' and two coeducational) were involved. The two questionnaires used in phase one were also used with these students and classroom observations, interviews of students, teachers, parents, parents and key informants were conducted. The data obtained from these interviews were analysed under a qualitative paradigm to probe deeper into issues related to teaching and learning of mathematics at the secondary level.

In an attempt to help in enhancing the mathematics achievement of students, a teaching and learning package was designed in the third phase of the study. This package was administered with 111 students from three secondary schools (one single boys', one single girls' and one coeducational). Lessons were prepared based on the ASEI/PDSI approach and taught for a period of three months using the strategies: cooperative learning, activity based and student centered teaching. The students were guided at different stages of a lesson to collaborate with their friends in the group to perform the activities involved and discover the underlying mathematical concepts. Proper scaffolding was provided to assist the students in their learning, and ample opportunities were provided to the students to be involved in critical thinking and to develop appropriate problem solving strategies. Drill exercises on the concepts learnt during the lessons were also given in the form of homework. Close monitoring of the students' progress was carried out during the class using a question-answer method, and class work set together with homework. A pre-test and post-test technique was used to evaluate cognitive gains that occurred during the period of the implementation phase. The WIHIC questionnaire also was used to analyse the perceptions of students on issues such as cooperation, their own involvement in the tasks and equity. Feedback from students in relation to their appreciation of the strategies used was collected in the form of reports. On the whole, the students were positive regarding the use of the strategies in motivating them to learn mathematics and helping them in understanding the underlying mathematical concepts.

The main emphasis in the third phase of the study was to enhance conceptual understanding in mathematics, and students' proficiency in the areas of conceptual understanding, procedural fluency and problem-solving skills were indeed improved. This outcome reflects the results of studies conducted by Rittle-Johnson and Stiegler (1998) who noted a high correlation between conceptual understanding and procedural skills. The enhancement of conceptual understanding was bound to bring about enhanced procedural skills which, as a result, better equipped the students to deal with problem solving. These outcomes could be noted throughout the implementation phase.

To discuss the impact of the teaching and learning package in the affective domain, analysis of data related to each of the scales of the WIHIC questionnaire — which was administered to the students towards the end of the implementation phase — is outlined next.

Student Cohesiveness: The mean for this scale was 4.02 with a standard deviation of 0.39. This showed the positive interrelationship that existed between the students, with peers helping out others in tasks. These interactions could be noted during the different group tasks and classroom discussions. Such a positive environment between students in the classes did help to enhance the teaching and learning process. The transcripts of students were analysed in Chapter Six and these provide evidence for the extent the students were enjoying the interrelationship which existed between them.

Teacher Support: The mean for this scale was 3.13 and the standard deviation was 0.47. This indicated that the students involved in the third phase of the study tended to be satisfied that personal attention to their difficulties and feelings were catered for, even though more personal and individual attention would have been desirable. One of the obstacles to this was the limited contact time with the students: just once a week and for forty minutes. Teachers should however note that personal interaction and involvement are important ingredients for providing a rich classroom environment and special effort should be dispensed to maximize these.

Involvement: The mean of 3.22 and standard deviation of 0.14 indicated that students were satisfactorily involved in classroom discussions. This showed the level of classroom interactions that were present during the lessons. Students were not passive and simply listening to the teacher but actively involved in meaningful conversations. Their ideas and suggestions were properly acknowledged and used during classroom discussions. They were also given the opportunities to explain the strategies they used to solve a particular mathematical problem.

Investigation: The investigative part of the lessons in mathematics was found to be similarly rated highly by the students with a mean of 3.22 and a standard deviation of 0.26. Students need to be involved in tasks where they make hypotheses, conduct experiments to investigate them and then draw conclusions. In this way their active involvement is ensured and the development of their conceptual understanding is enhanced. Evidence of the positive attitude of students towards the activities carried out during the classes and their usefulness in enhancing the knowledge of mathematical concepts was described in Chapter Six.

Task Orientation: This was the scale with the highest mean score (4.18) in the WIHIC questionnaire, with a standard deviation of 0.29. This suggests the high level of commitment and dedication of the students to the lessons, and tends to support the use of the strategies that were designed for the teaching and learning package.

Cooperation: A mean score of 3.63 with a standard deviation of 0.24 indicated that students were satisfactorily involved in a cooperative environment. Evidence of the students' positive attitude towards cooperation and the extent to which this helped them in their learning of mathematics were provided through the transcripts discussed in Chapter Six. It should be mentioned that the students were normally used to be working individually for so many years, and cooperative learning strategies were brought in only when the implementation stage started. They became used to the strategies without much delay and I believe that they will derive more from the benefits of cooperative learning given more time.

Equity: The principle of equity is important in one's life and the mean score of 3.77 with a standard deviation of 0.16 indicates that the students believed that the teacher treated students equally. This belief should be consolidated with more equitable acts and behaviour in the classroom environment of the students.

Having briefly discussed the study and the improvement of students both in the cognitive and affective domains, the main findings of the study are now presented in the following section.

Main findings of the study

One of the main outcomes of the study is the confirmation it has provided that Mauritian boys and girls are performing differently in mathematics at the secondary level, with boys outperforming girls not only in the overall mathematics test but also in the strands Number, Algebra, Geometry and Statistics. These differing performances were discussed in detail in Chapter Four. The findings of boys outperforming girls in mathematics at secondary level agree with those of Githua and Mwangi (2003) in Kenya, Afrassa (2002) in Ethiopia and other parts of Africa (Kogolla, Kisaka, & Waititu, 2004). Similar findings were also reported in other studies (Koller, Baumert, & Schnabel, 2001). The findings, however, contradict those of Hanna (2003), Boaler (1997) and Vale, Forgasz, & Horne (2004). It seems that the Western World has successfully tackled the problem of girls' underachievement in mathematics as opposed to boys through different intervention programmes, while in the developing countries it still remains to be done.

This study also aimed to determine the factors that impact on the mathematics achievement of boys and girls in mathematics. Research Question One in fact was devoted to identify such factors. The contribution of each factor identified was discussed in Chapter Five when the Research Question One was answered. One of the findings of the present study relates to the method of teaching normally employed in the mathematics classrooms at secondary level in Mauritius. The method was teacher-centered, and students were passive and on the receiving end, learning algorithms to apply to solve mathematical problems. This phenomenon reflected the lessons described by Nunes and Bryant (Nunes & Bryant, 1997), and the descriptions of primary classes in Mauritius (Griffiths, 1998, 2000, 2002). It seems that insufficient opportunities are provided for students to be involved in their own learning, and emphasises the algorithmic procedures used for solving mathematics problems. It seems that the examination-driven curriculum in Mauritius leads to a more teacher-centered curriculum.

Teachers were found to be playing a fundamental role in influencing students' learning of mathematics, as noted by Hanna & Nyhof-Young (1995). They also helped students to develop a positive attitude towards mathematics and motivate them towards the subject. The respect students have for their teachers could be noted during the classroom observations and interviews. This supports the finding of Aldridge, Fraser and Huang (Aldridge, Fraser, & Huang, 1999) concerning the respect students had for their teacher in Taiwan. It was also found that teachers were seen to be strict, and that students appreciated the strictness, claiming that this helped them to have a disciplined class in which to learn mathematics. Evidence of this can be found in the transcripts of students' interviews which were discussed in Chapter Five. Teachers were found to be acting as role models, were possessing sound leadership skills and were of a helpful nature. However, there were teachers who had a gender bias in their own perception. As described by Elwood (Elwood, 1999), they tended to describe male students as able in mathematics and female students as being uncertain and not possessing enough faith in their own ability. These findings were more common for average and low performing girls — findings which are in agreement with those of Tiedemann (2000). Cases where negative messages were sent to girls about their performance in mathematics by teachers were noted in the present study also.

Interestingly, parental interest and involvement in their children's education is high in Mauritius. The contributions of parents towards the children's learning in mathematics were discussed in Chapter Five. It was found that students are conscious of their parental aspirations and this plays an important role in their motivation towards education. It should be noted that parents' support towards education in Mauritius is no longer gender-biased now – as it used to be. Parents believe in the power of education and the success of their children depends to a great extent on their educational success. However, the way of attributing success and failure in mathematics to boys and girls still followed the pattern as described by Raty et al. (2002) where the success of boys was attributed to talent, while the success of girls was due more to effort.

Peers were found to be influential in a child's learning of mathematics and, in some cases, in decisions to proceed further with other mathematical courses and the learning of mathematics in general. This agrees to the findings of Opdenakker & Van Damme (2001), Sam & Ernest (1999) and Hoxby (2002). Peer influence is not restricted to the classroom only or to school mates, but from a much larger group through private tuition. The practice of private tuition allows students of different regions, colleges, cultures and social classes to be together and consequently to form a larger peer group. This study was restricted to the peer influence within the classroom towards the teaching and learning of mathematics.

A correlation coefficient of 0.336 between attitude towards mathematics and performance in the mathematics test was noted in this study. However, no gender difference in attitude towards mathematics was observed. A positive attitude towards mathematics and interest in the subject tends to motivate students into putting more effort into the subject, and this consequently enhanced their mathematics achievement. Concerning success or failure in mathematics, it was found that students attributed success primarily to effort, and failure to lack of effort — evidence coming from the transcripts of students' interviews as discussed in detail in Chapter Five. These findings agreed with the findings of Mooney and Thornton (1999) but no apparent gender differences were noted — contradicting the outcomes reported by Ernest (1994) and Leder, Forgasz and Solar (1996). It can be deduced that Mauritian girls are different to Australian and English girls in this respect.

Prior ability in mathematics was found to play an important role in the mathematics achievement of students as claimed by O'Conner and Miranda (2002). This is so because of the hierarchical nature of the subject — mathematical concepts build on prior ones. This finding proved to be important as the way mathematics is being taught at primary and lower secondary levels should be taken into account. There are cases of schools in Mauritius where inexperienced teachers are being sent to lower secondary classes and the more qualified and experienced ones deal only with upper classes. The mathematical concepts have to be learnt properly right from lower classes to ensure a solid base for the students to assist them in their learning of mathematics at each successive level.

A summary of these factors that impact on the mathematics performance of boys and girls in Mauritius as identified through the present study is provided in Figure 7.1 on Page 223.

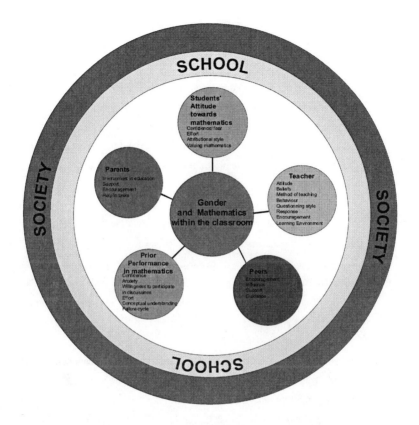

Figure. 7.1: Model for factors impacting on gender and mathematics within the classroom

Note that only the main factors having a direct impact within the classroom have been described in this figure. The figure also includes two outer concentric circles, which show that there are additional school and societal factors which, in someway or the other, have an impact on gender and mathematics.

Another factor, language, was found to also play a major role in the teaching and learning of mathematics. It was revealed in this study that students were having problems tackling word problems or problems related to application to real life situations. Similar outcomes were highlighted in a study conducted by Zevenbergen (2001). Indeed, there is considerable debate related to the issue of language and education in Mauritius and a pilot project is to be conducted by the Government in

2006 where the local language (Creole) will be used as the medium of instruction at the primary level.

It was found in this study that parents, in general, believed that private tuition would help their children in their learning of mathematics. While agreeing that this extra coaching had considerable financial implications, parents felt that providing private tuition was part of the support that they could extend to their children. It should be noted that the issue of private tuition is a significant one for the Mauritian society, as discussed in Chapter One.

The present study also found that students from the Chinese community tend to perform well in mathematics in general. This was discussed in Chapter Four when the regression analysis was conducted, and again in Chapter Five. This finding agrees with research conducted for the TIMSS and also with studies by Rao, Moely and Sachs (2000). Gender differences within the community could not be analysed in this present study.

Another finding of this study is that the mathematics performances of students in the different zones in Mauritius are significantly different, which agrees with the findings of the Monitoring Learning Achievement Study (2003). The zoning system was introduced in Mauritius in 2002 for the admission to secondary schools after the CPE examinations. The study conducted at the primary level (Monitoring Learning Achievement, 2003) and the present study's results tend to show that students in the different zones are performing significantly differently in mathematics.

This study has also found out that students perceived mathematics to be a masculine pursuit and that they have a stereotyped image of mathematicians and mathematics. The drawings students made for a mathematician corresponded to studies conducted by Sumida (2002). From some of the drawings it could be deduced that the students thought that mathematics was much beyond their ability, as the mathematicians was shown with great intellectual powers. Teachers need to determine their students' perceptions towards the subject as negative perceptions may influence the student's involvement and subsequent achievement in that subject. This is discussed in more detail in a later section.

Implications for teaching and learning

The most direct impact of this study will hopefully be in the classroom and will help teachers to use the findings, in particular:

-   using student-centered teaching approaches
-   using meaningful activities in their classrooms
-   promoting conceptual understanding in mathematics
-   emphasising process rather than product during problem-solving sessions
-   promoting collaborative learning in mathematics classes
-   helping students to develop a positive attitude towards mathematics
-   motivating students in their learning of mathematics
-   creating a conducive environment for learning of mathematics
-   enhancing the mathematics achievement of all students
-   promoting equity in education.

Each is explained in detail later in this chapter.

Teachers will have evidence on how different strategies can be incorporated with success into their regular classroom transactions and within their schedule of work. One teacher, who acted as a member check in the third phase of the study, stated that using cooperative learning and student-centered methods would be very time consuming and that teachers would face difficulties in completing the syllabus (See Appendix Eighteen for transcript of the feedback). As argued in the previous chapters, one of the main worries of teachers, rectors, parents, and inspectors is that the syllabus should be thoroughly completed. All that is required is a readjustment of teachers' time schedule and advanced planning of the lessons. It is hoped that the lessons described in the teaching and learning package will provide teachers with ideas on how to plan lessons where the teaching will be student-centered, activity-based and use collaborative learning. These ideas will be used in the teacher training programmes both for pre-service and in-service teachers. Workshops to discuss these issues will also be conducted on a regional basis.

This study should also help teachers and other interested parties to realize how important they are in the educational lives of students. By reading the transcripts of the interviews of the students and other studies in the literature search, they will realize how they can influence students. Teachers should think on the ways they react to the students, to boys and girls and how they may help in achieving gender equity.

This study has shown that it is imperative that mathematics teacher be aware of the perceptions that students hold concerning the subject, and this is especially so for mathematics at secondary school. The practitioner should teach in a way that helps all students develop a positive attitude towards mathematics and enhances their performance in the subject. Teacher behaviour, and the instructional strategies that they use, affect students' skills, interests, and retention rates in mathematics and science (Kahle, 1989). It is likely that when the emphasis in the classroom is placed more on the outcomes of learning rather than on the process of learning, students are unable to develop mathematical concepts in a consistent and fruitful way.

Teachers also need to be careful about the pace with which they demonstrate mathematical techniques to students. Students should be given time to understand the processes involved. At times, students see themselves far from the subject. Indeed, Picker & Berry (2000) highlighted the fact that "there appears to be no other subject studied in school where pupils are placed at such a distance from the discipline than occurs in mathematics" (p. 90). It is imperative that teachers devise ways and means to improve the image of mathematicians (and particularly female ones) effectively in their classroom.

A cycle has been proposed by Picker and Berry (2000) regarding the perpetuation of stereotypical images of mathematicians and mathematics and is illustrated in Figure 7.2. Commencing at the base of the figure, the cycle describes how a student may develop a negative attitude towards mathematics and a consequent belief system.

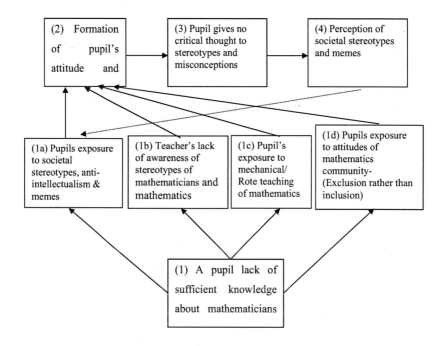

Figure 7.2: A Cycle That Perpetuates Stereotypes (Picker and Berry, 2000, p. 86)

If the belief system is not questioned, students' stereotypes about mathematicians and mathematics will be maintained. The stereotyped views will be then be exchanged with other students. Negative messages such as: *mathematics is boring, mathematics was the most difficult subject I ever had, I was never good at maths* ... will be conveyed to peers and even younger children. It should be highlighted that such statements can often be heard in conversations, but it is rarer to hear *I was not good at reading.* Mathematics is a particular subject where negative statements are common and are accepted in society.

A report by the National Research Council (1989) included the following comments concerning the state of the image of mathematics held by students.

> Unfortunately, as children become socialised by schools and society, they begin to view mathematics as a rigid system of externally dictated rules

governed by standards of accuracy, speed and memory. Their view of mathematics shifts gradually from enthusiasm to apprehension, from confidence to fear. Eventually, most students leave mathematics under duress, convinced that only geniuses can learn it (cited in Picker & Berry, 2000, p. 67).

Teachers at the secondary level should endeavour to challenge students' stereotypical views of mathematicians and mathematics to empower them in their learning of the subject.

Since the implementation of the teaching and learning package produced positive outcomes both in cognitive as well as affective domains, I recommend that the strategies used in this study be implemented in classroom practices. In short, the evidence is here that the following suggestions will help the practitioners in the teaching and learning process of mathematics at the secondary level, improve the mathematics achievement of students and achieve gender equity.

- **Proper planning of lessons is very important. The way the lesson should commence, the questions that would establish the start up, the activities to be carried out, the teaching aids needed and probable misconceptions students may have prior to the lesson should be carefully considered beforehand.**
- **Cooperative learning should be encouraged in mathematics classes. Teachers need to assist the students to work within a group. Simply placing them together around a table will not ensure cooperative learning. Students should be helped in that process and the teacher should know his/her role in that setting.**
- **Emphasis should be laid more on process rather than product as advocated by Dekker and Elshout-Mohr (2004). Teaching should be carried out in view of promoting the conceptual understanding of students and this will result in the development of procedural and problem solving skills.**

- Students should be the focus in the classrooms and their involvement in well designed activities catered for. The teaching and learning pace should be chosen appropriately to match students' learning development to engage them in class discussions. Equal attention should be given to boys and girls. All the students should be given ample opportunities to participate in the classroom discussions. Appropriate prompts and wait time need to be given to boys and girls to engage them in critical thinking. Any prior negative image of mathematics or negative attitude towards mathematics for boys and girls should be noted and treated properly to help each and every student develop a liking of the subject and perform better.

- A mathematics laboratory needs to be set up in schools where different posters with short summaries can be displayed. A laboratory, with an array of mathematical models and tools, will provide an appropriate learning environment for the effective teaching and learning of mathematics.

- Regular meetings with parents need to be conducted to ensure their collaboration and support in the educational lives of their children. Advice and support could be provided in needy cases.

- As a foundation member of the Mathematics Teachers Association in Mauritius, I believe that the association should be proactive to provide a platform for teachers in Mauritius to share their difficulties, experiences and know-how to help each other in their professional jobs.

- Regular in-service sessions should be run by the Mauritius Institute of Education and other sister institutions in the educational field to discuss latest developments in the field of education. It was explained while answering Research Question One that pre-service teacher training is not compulsory at the secondary level in Mauritius and teachers undergo professional development involving pedagogical issues while they are full- fledged teachers. It should also be pointed out that teachers who have followed pedagogical courses at the Mauritius Institute of Education have, till recently, rarely had the chance to enroll in other courses which would have further helped them in their professional development. Consequently, the teachers tend to be unaware of new developments in the educational field which they could implement at the classroom level. Regular in-service sessions on mathematics education and other relevant educational issues should help teachers with their professional activities.

- Changes should be brought in the policy dealing with teacher training and in the programmes offered in the teacher training courses. Pre-service training should be made compulsory at the secondary level in Mauritius where teachers are equipped with strategies for effective teaching and evaluation and where they have discussed issues related to equity in education.

Limitations of the study

Any educational research is conducted in a certain frame of reference and this inevitable brings forward some restrictions. The limitations that were encountered during the study follow:

Sample selection

For each of the three phases, different samples of schools and students were chosen. Schools were grouped into the different types (as described in Chapter Three) and the schools were selected in such a way to reflect the proportion that existed in the population. Moreover, categories of high achieving, average and low achieving students were also taken into consideration as per the performance of the schools in the School Certificate examinations. It was, however, assumed that students in the same category were comparable across schools in terms of mathematics achievement.

Data collection

In the first phase of the study I ensured that I visited each of the schools in the sample and discussed the objectives of the study as well as the instruments that were used. However, it was the classroom teacher who administered the test and the questionnaire related to the attitude of students towards mathematics. It was assumed that all the teachers explained the objectives of the test to the students and the students in all the schools took the test with the same degree of seriousness. Concerning data collection through classroom observations, field notes were taken based on a previously designed observational schedule. However, it may have been that certain events that were occurring in the classrooms missed my attention. Furthermore, only audio taping was used for the interviews conducted. The transcripts of the interviews were then used for data analysis. If the interviews were videotaped, the facial expressions of the interviewees could be seen and taken into consideration while analyzing the data. The use of video recording could have helped in the same way for classroom observations.

Instrumentation

One of the main instruments used in this study was the mathematics questionnaire which was self-designed. Though its 'face validity' was ensured by discussing with colleagues in the department and an expert in the educational field, and the reliability of the results obtained was satisfactory, an instrument which had been validated and used in other studies may have been more appropriate to use.

Procedural

It should be mentioned that the culture of the secondary school educational system in Mauritius is such that the students are not used to filling in questionnaires in educational research, not are they used to being asked about their perceptions on issues pertaining to teaching and learning. However, the results concerning the reliability coefficients of each of the scales in the QTI and WIHIC questionnaires proved that this cultural limitation did not appear to pose much of a problem for the students — either to be involved in such a process or understanding the questions asked.

Time also was a major constraint as the implementation phase was held over a period of three months. Moreover, classes in each school were conducted once per week for a period of forty minutes. This limited time restricted the activities that could be carried out in the class. Moreover, there was little choice in relation to the topics that were dealt with in the classes. I was required to fit in and conform with the scheme of work, and work on the topics that were being studied at that time.

Another possible limitation was that the students were exposed to two different methods of mathematics teaching — one where they were required to take the main roles in their own learning (once a week), and the other more traditional way where they were dependent on the teacher (for the remainder of the week).

Yet another factor which could have affected the results in the pre-test and post-test to some extent was that the scores were not to be counted for the students' first term results. Though the objectives of the exercise were clearly explained to students and their total involvement requested, some may not performed to the best of their ability.

Finally it should be pointed out that the teaching aids used while conducting the classes were limited to locally available resources, and ICT techniques were not exploited. In Mauritius, Computer Studies has been taught as a subject for about fifteen years, but its application in other subjects has been almost nonexistent. Many forums have been organized calling for the effective use of ICT within subject areas, but the lack of facilities and resources in the secondary schools have delayed such an initiative.

Recommendations for further research

To my knowledge, this study was the first where questionnaires were used in Mauritius to assess students' perceptions of their learning environment related to their classrooms, with the reliability of the results obtained comparable to studies conducted in other parts of the world. This should encourage further studies where other aspects of the learning environment could be investigated at primary, secondary and tertiary levels.

Further studies on gender and mathematics at secondary level should be conducted in relation to single sex and coeducational schools. An investigation of the attitudes towards mathematics and the performance of boys and girls in single sex schools, as compared to those in coeducational schools, could prove to be important also.

This study has just touched upon the relationship between culture and performance in mathematics. Mauritius is a multicultural country with a blend of different cultures, and an in-depth study wherein the issue of gender and mathematics in relation to ethnicity is examined would be valuable.

The study also focused on strategies that could be employed within the classroom to help students enhance their performance in mathematics. Further studies can be carried out to identify other strategies that will assist in enhancing students' attitude towards mathematics and their mathematical achievement.

As this study was conducted in three phases with a different sample of students for each phase, a longitudinal study could be carried out, where the same batch of students from Form III is targeted and followed throughout until they reach Form V. The attitudes of students towards mathematics, performance in the subject and perceptions in other related issues could be studied and explored and any change in these factors over time could be investigated.

Concluding words

It is hoped that this study will sensitize teachers, rectors, inspectors, curriculum developers, parents, policy makers and other stakeholders in the educational field of this important issue of gender and mathematics and how boys and girls in Mauritius are performing in the subject. The study will hopefully bring to the attention of those people who are involved in education the voices of students, teachers, parents and key informants on issues related to the teaching and learning of mathematics. All partners in the educational field should endeavour to work hand in hand to devise ways and means to help the children of Mauritius perform better in mathematics, and in education in general, to meet the challenges of the future. Hopefully the study has provided some guidance towards achieving this goal.

References

Afrassa, T. M. (2002). *Changes in mathematics achievement over time in Australia and Ethiopia.* Adelaide, Australia: Shannon Research Press.

Alderman, M. K. (1999). *Motivation for achievement: Possibilities for teaching and learning.* Mahwah, New Jersey: Lawrence Erlbaum Associates.

Aldridge, J. M., Fraser, B. J., & Huang, T. C. I. (1999). Investigating classroom environments in Taiwan and Australia with multiple research methods. *Journal of Educational Research, 93,* 48-57.

Alkhateeb, H. M. (2001). Gender differences in mathematics achievement among high school students in the United Arab Emirates, 1991-2000. *School Science and Mathematics, 101*(1).

American Association of University Women Educational Foundation. (1998). *Gender gaps: Where schools still fail our children.* Washington, D.C.

Armstrong, J. M., & Price, R. A. (1982). Correlates and predictors of women's mathematics participation. *Journal for Research in Mathematics Education, 13,* 99-109.

Aronson, E., Wilson, T. D., & Akert, R. M. (1997). *Social psychology* (2nd ed.). New York: Longman.

Atweh, B., Cooper, T., & Kanes, C. (1992). The social and cultural context of mathematics education. In B. Atweh & J. Watson (Eds.), *Research in mathematics education in Australasia* (pp. 43-66). Brisbane: Mathematics Education Research Group.

Barwell, R. (2005). Working on arithmetic word problems when English is an additional language. *British Educational Research Journal, 31*(3), 329-348.

Battista, M. T. (1990). Spatial visualization and gender differences in school geometry. *Journal for Research in Mathematics Education, 21*(1), 47-60.

Becker, J. R. (2003). Gender and mathematics: An issue for the twenty-first century. *Teaching Children Mathematics, 9*(8).

Beghetto, R. A. (2004). *Toward a more complete picture of student learning: Assessing students' motivational beliefs.* Retrieved 20 May 2005, from http:PAREonline.net/gtvn.asp?v=9&n=15

Bessoondyal, H., & Fisher, D. L. (2003). Assessing the classroom learning environment in a teacher training institution: A case study. In D. L. Fisher & T. Marsh (Eds.), *Making Science, Mathematics and Technology Education accessible to All, Proceedings of the Third Conference on Science, Mathematics and Technology Education, South Africa* (Vol. 2, pp. 447-454). Perth: Curtin University of Technology.

Boaler, J. (1997). Reclaiming school mathematics: The girls fight back. *Gender and Education, 9*(3), 285-306.

Boaler, J. (1998). Open and closed mathematics: Student experiences and understandings. *Journal for Research in Mathematics Education, 29*(1), 41-62.

Bong, M. (2005). Within-grade changes in Korean girls' motivation and perceptions of the learning environment across domains and achievement levels. *Journal of Educational Psychology, 97*(4), 656-672.

Bunwaree, S. (1994). *Mauritian education in a global economy*. Rose-Hill, Mauritius: Edition de l'Ocean Indien Ltee.

Bunwaree, S. (1996). *Gender, education/training and development in Mauritius*: UNDP.

Bunwaree, S. (2002). Introduction and overview. In S. Bunwaree (Ed.), *Rethinking development: Education & inequality in Mauritius* (pp. 1-11). Reduit, Mauritius: Centre for Educational Research and Publications, MIE.

Burton, L. (1986). *Girls into maths can go*. London: Holt, Rinehart & Winston.

Burton, L. (Ed.). (1990). *Gender and mathematics: An international perspective*. London: Cassell.

Cahill, L. (2005, 25 April). His brain, her brain. *Scientific American*.

Cameron, J., Pierce, W. D., Banko, K. M., & Gear, A. (2005). Achievement-based rewards and intrinsic motivation: A test of cognitive mediators. *Journal of Educational Psychology, 97*(4), 641-655.

Carey, D. A., Fennema, E., Carpenter, T. P., & Franke, M. L. (1995). Equity and mathematics education. In W. G. Secada, E. Fennema & L. B. Adajian (Eds.), *New directions for equity in mathematics education* (pp. 93-125). New York: Cambridge University Press.

Carr, M., Jessup, D., & Fuller, D. (1999). Gender differences in first-grade mathematics strategy use: parent and teacher contributions. *Journal for Research in Mathematics Education, 30*(1), 20-46.

Casey, M. B., Nuttall, R. L., & Pezaris, E. (2001). Spatial- mechanical reasoning skills versus mathematics self - confidence as mediators of gender differences on mathematics subtests using cross-national gender-based items. *Journal for Research in Mathematics Education, 32*(1), 28-57.

Caunhye, N. (1993). *Women's education in Mauritius from 1948 to 1976.* University of Mauritius, Reduit, Mauritius.

Chambers, D. W. (1983). Stereotypic images of the scientist: The Draw-A-Scientist test. *Science Education, 67*, 255-256.

Chionh, Y. H., & Fraser, B. J. (1998). *Validation and use of the What is Happening in this Class (WIHIC) questionnaire in Singapore.* Paper presented at the annual meeting of the American Educational Research Association, San Diego, CA.

Clarkson, P. C. (1992). Language and mathematics: A comparison of bilingual and monolingual students of mathematics. *Educational Studies in Mathematics, 23*(4), 417-429.

Clarkson, P. C., & Galbraith, P. (1992). Bilingualism and mathematics learning: Another perspective. *Journal for Research in Mathematics Education, 23*(1), 34-44.

Cockcroft, W. H. (1982). *Mathematics counts.* London: Her Majesty's Stationery Office.

Cohen, L., Manion, L., & Morrison, K. (2000). *Research methods in education* (5 ed.). London: Routledge Falmer.

Cuevas, G., J. (1984). Mathematics learning in English as a second language. *Journal for Research in Mathematics Education, 15*(2), 134-144.

Davidson, N. (1996). Small-group cooperative learning in mathematics. In T. J. Cooney & C. R. Hirsch (Eds.), *Teaching and learning mathematics in the 1990's* (pp. 52-61). Reston, VA: National Council of Teachers of Mathematics.

Davidson, N. (Ed.). (1990). *Cooperative learning in mathematics: A handbook for teachers.* New York: Addison-Wesley Publishing Company.

DeBoer, G., Morris, K., Roseman, J. E., Wilson, L., Caprano, M. M., Caprano, R., et al. (2004). *Research issues in the improvement of mathematics teaching and learning through professional development.* Retrieved 25 May 2005, from http://www.project2061.org/publications/articles/IERI/AERA2004.pdf

Dekker, R., & Elshout-Mohr, M. (2004). Teacher interventions aimed at mathematical level raising through collaborative learning. *Educational Studies in Mathematics, 56,* 39-65.

Denscombe, M. (1998). *The good research guide.* Buckingham: Open University Press.

Denzin, N. K., & Lincoln, Y. S. (Eds.). (2000). *Handbook of qualitative research.* Thousand Oaks: Sage Publications.

Digest of Educational Statistics. (2003). Central Statistics Office, Ministry of Finance and Economic Development.

Dunne, M., & Johnson, J. (1994). Research in gender and mathematics education: The production of difference. In P. Ernest (Ed.), *Mathematics, education and philosophy: An international perspective* (pp. 221-229). London: The Falmer Press.

Eccles, J. S., & Jacobs, J. E. (1986). Social forces shape math attitudes and performance. *Signs, 11,* 367-380.

Ellerton, N. F., & Clarkson, P. C. (1996). Language factors in mathematics teaching and learning. In A. Bishop, J, K. Clements, C. Keitel, J. Kilpatrick & C. Laborde (Eds.), *International Handbook of Mathematics Education.* Netherlands: Kluwer Academic Publishers.

Elliot, P. C., & Kenney, P. (1996). *Communication in mathematics, K-12 and beyond.* Reston, VA: National Council of Teachers of Mathematics.

Elwood, J. (1999). Gender, achievement and the 'gold standard': Differential performance in the GCE A level examination. *The Curriculum Journal, 10*(2), 189-208.

Ercikan, K., McCreith, T., & Lapointe, V. (2005). Factors associated with mathematics achievement and participation in advanced mathematics courses: An examination of gender differences from an international perspective. *School Science and Mathematics, 105,* 5-14.

Ernest, P. (1994a). *The nature of mathematics and equal opportunities.* Exeter: University of Exeter School of Education.

Ernest, P. (1994b). *Psychology of learning mathematics.* Exeter: University of Exeter School of Education.

Ethington, C. A. (1990). Gender differences in mathematics: An international perspective. *Journal for Research in Mathematics Education, 21,* 74-80.

Ethington, C. A. (1992). Gender differences in a psychological model of mathematics achievement. *Journal for Research in Mathematics Education, 23*(2), 166-181.

Fennema, E. (1974). Mathematics learning and the sexes: a review. *Journal for research in mathematics education, 5*(3), 126-139.

Fennema, E. (1990). Justice, equity, and mathematics education. In E. Fennema & G. C. Leder (Eds.), *Mathematics and gender* (pp. 1-9). New York: Teachers College Press.

Fennema, E. (1995). Mathematics, gender and research. In B. Grevholm & G. Hanna (Eds.), *Gender and mathematics education* (pp. 21-38). Sweden: Lund University Press.

Fennema, E. (1996). Mathematics, gender, and research. In G. Hanna (Ed.), *Towards gender equity in mathematics education* (pp. 9-26). Netherlands: Kluwer Academic Press.

Fennema, E. (2000). *Gender and mathematics: what is known and what do I wish was known.* Retrieved 28 August 2002, from http://www.wcer.wisc.edu/nise/News_Activities/Forums/Fennemapaper.htm

Fennema, E., Carpenter, T. P., Jacobs, V. R., Franke, M. L., & Levi, L. W. (1998). A longitudinal study of gender differences in young children's mathematical thinking. *Educational Researcher, 27*(5), 6-11.

Fennema, E., & Leder, G. C. (Eds.). (1990). *Mathematics and gender.* New York: Teachers College Press.

Fennema, E., & Sherman, J. (1977). Sex-related differences in mathematics achievement, spatial visualization, and sociocultural factors. *American Educational Research Journal, 14,* 51-71.

Fennema, E., & Tartre, L. A. (1985). The use of spatial visualization in mathematics by boys and girls. *Journal for Research in Mathematics Education, 16,* 184-206.

Fisher, D. L., Rickards, T., & Fraser, B. J. (1996). Assessing teacher-student interpersonal relationships in science classes. *Australian Science Teachers Journal, 42*(3), 28-33.

Forbes, S. D. (1999). *Measuring students' education outcomes: Sex and ethnic differences in mathematics.* Unpublished doctoral dissertation, Curtin University of Technology, Perth, Western Australia.

Forgasz, H. J., Leder, G. C., & Gardner, P. L. (1999). The Fennema - Sherman Mathematics as a Male Domain Scale Reexamined. *Journal of Research in Mathematics Education, 30*(3), 342-348.

Fraenkel, J. R., & Wallen, N. E. (1993). *How to design and evaluate research in education*: McGraw Hill.

Frankenstein, M. (1995). Equity in the mathematics classroom: Class in the world outside the class. In W. G. Secada, E. Fennema & L. B. Adajian (Eds.), *New directions for equity in mathematics education* (pp. 165-190). Cambridge: Cambridge University Press.

Fraser, B. J. (1998). Science learning environments: Assessment, effects and determinants. In B. J. Fraser & K. G. Tobin (Eds.), *International handbook of science education* (pp. 527-564). Dordrecht, The Netherlands: Kluwer Academic Publishers.

Friedmann, L. (1989). Mathematics and the gender gap: A meta-analysis of recent studies on sex differences in mathematical tasks. *Review of Educational Research, 59,* 185-213.

Fullarton, S. (1993). *Confidence in mathematics: The effects of gender*: Deakin University. National Centre for Research and Development in Mathematics Education.

Gallagher, A., & De Lisi, R. (1994). Gender differences in scholastic aptitude test - mathematics problem solving among high-ability students. *Journal of Educational Psychology, 86*(2), 204-211.

Gardner, H. (1980). *Artful scribbles: The significance of children's drawings.* New York: Basic Books.

Gardner, H. (1993). *Frames of mind: The theory of multiple intelligences.* New York: Basic Books.

Gipps, C., & Murphy, P. (1994). *A fair test? : assessment, achievement and equity.* Buckhingham: Open University.

Githua, B. N., & Mwangi, J. G. (2003). Students' mathematics self-concept and motivation to learn mathematics:relationship and gender differences among Kenya's secondary-school students in Nairobi and Rift Valley provinces. *International Journal of Educational Development, 23*(5), 487-499.

Goh, S. C., & Fraser, B. J. (1995). *Learning environment and student outcomes in primary mathematics classrooms in Singapore.* Paper presented at the annual meeting of the American Educational Research Association, San Francisco, CA.

Gomez- Chacon, I. M. (2000). Affective influences in the knowledge of mathematics. *Educational Studies in Mathematics, 43*, 149-168.

Goodell, J. E., & Parker, L. H. (2001). Creating a connected, equitable mathematics classroom: Facilitating gender equity. In B. Atweh, H. Forgasz & B. Nebres (Eds.), *Sociocultural research on mathematics education: An international perspective* (pp. 411-431). Mahwah, NJ: Lawrence Erlbaum Associates.

Goodell, J. E., & Parker, L. H. (2003). *Equity in mathematics education: Characteristics of a connected, equitable classroom.* Paper presented at the Third Conference on Science, Mathematics and Technology Education, South Africa.

Goodenough, F. L. (1926). *Measurement of intelligence by drawings.* New York: Harcourt Brace.

Goodenow, J. (1977). *Children's drawing.* London: Open Books.

Goodoory, C. (1985). *A survey of private at Form V level in secondary schools in Mauritius.* Unpublished dissertation for Post Graduate Certificate in Education, Mauritius Institute of Education.

Griffiths, M. (1998). *Stakeholders voices: A socio-cultural approach to describing and extending an understanding of primary education in Mauritius.* Unpublished PhD thesis, Edith Cowan University.

Griffiths, M. (2000). Learning for All? Interrogating children's experiences of primary schooling in Mauritius. *Teaching and Teacher Education*(16), 785-800.

Griffiths, M. (2002). Equality of opportunity in Mauritian primary education. In S. Bunwaree (Ed.), *Rethinking development: Education & inequality in Mauritius* (pp. 27-48). Reduit, Mauritius: Centre for Educational Research and Publications, MIE.

Grouws, D. A., Cooney, T. J., & Jones, D. (Eds.). (1989). *Perspectives on research on effective mathematics teaching, Volume 1*. Reston, Virginia: National Council of Teachers of Mathematics.

Guba, E. G., & Lincoln, Y. S. (1989). *Fourth generation evaluation*. London: Sage Publications.

Gutbezahl, J. (1995). *How negative expectancies and attitude undermine females' math confidence an performance: a review of the literature*. Retrieved 27 December, 2002, from http://www.fin.org/jennyg/articles/litreview.html

Hammrich, P. L. (2002). *Gender equity in science and mathematics education: Barriers of the mind*. Retrieved 4 June 2005, from http://www.rbs.org/currents/0601/gender_equity.shtl

Hanna, G. (1986). Sex differences in mathematics achievement of eight graders in Ontario. *Journal for Research in Mathematics Education, 17*, 231-237.

Hanna, G. (2003). Reaching gender equity in mathematics education. *The Educational Forum, 67*(3), 204-214.

Hanna, G. (Ed.). (1996). *Towards gender equity in mathematics education*. Dordretch: Kluwer Academic Press.

Hanna, G., Kundiger, E., & Larouche, C. (1990). Mathematical achievement of grade 12 girls in fifteen countries. In L. Burton (Ed.), *Gender and mathematics: An international perspective* (pp. 87-97). London: Cassell.

Hanna, G., & Nyhof-Young, J. (1995). An ICMI study on gender and mathematics education: Key issues and questions. In B. Grevholm & G. Hanna (Eds.), *Gender and mathematics education* (pp. 7-14). Sweden: Lund University Press.

Harris, D. B. (1963). *Children's drawings as measures of intellectual maturity: A revision and extension of the Goodenough Draw-A-Man Test*. New York: Harcourt Brace Jovanovich, Inc.

Hiebert, J., & Carpenter, T. P. (1992). Learning and teaching with understanding. In D. A. Grouws (Ed.), *Handbook of research on mathematics teaching and learning* (pp. 65-97). New York: Macmillan.

Hoxby, C. M. (2002). *The power of peers*. Retrieved 24 May 2005, from http://www.org/20022/56.html

Hyde, J. S., Fennema, E., & Lamon, S. J. (1990). Gender differences in mathematics performance: A meta-analysis. *Psychological Bulletin, 107,* 139-155.

Hyde, J. S., & Jaffe, S. (1998). Perspectives from social and feminist psychology. *Educational Researcher, 27*(5), 14-16.

Jaworski, B. (1994). *Investigating mathematics teaching: A constructivist enquiry.* London: Falmer Press.

Jaworski, B. (2003). Research practice into/influencing mathematics teaching and learning development: Towards a theoretical framework based on co-learning partnerships. *Educational Studies in Mathematics, 54,* 249-282.

Johnson, D. W., & Johnson, R. T. (1990). Using cooperative learning in mathematics. In N. Davidson (Ed.), *Cooperative learning in mathematics: A handbook for teachers* (pp. 103-124). New York: Addison-Wesley Publishing Company.

Kahle, J. B. (1989). Image of Scientists: Gender issues in science classrooms, *What Research Says to the Science and Mathematics Teacher.* Perth: The Key Centre for School Science and Mathematics, Curtin University of Technology.

Kaphesi, E. (2003). The influence of language policy in education on mathematics classroom discourse in Malawi: The teachers' perspective. *Teacher Development, 7*(2), 265-285.

Kastberg, S. G., D'Ambrosio, B., McDermott, G., & Saada, N. (2005). Context matters in assessing students' mathematical power. *For the Learning of Mathematics, 25*(2), 10-15.

Khalid, M. (2004). *Enhancing the mathematical achievement of technical education students in Brunei Darussalam using a teaching and learning package.* Unpublished PhD thesis, Curtin University of Technology, Perth, Australia.

Khisty, L. L., & Chval, K. B. (2002). Pegagogic discourse and equity in mathematics: when teachers' talk matters. *Mathematics Education Research Journal, 14*(3).

Kloosterman, P. (1990). Attributions, performance following failure, and motivation in mathematics. In E. Fennema & G. C. Leder (Eds.), *Mathematics and gender* (pp. 96-127). New York: Teachers College Press.

Knapp, M. S., Shields, P. M., & Turnbull, B. J. (1995). Academic challenge in high-poverty classrooms. *Phi Delta Kappan, 76*(10), 70-76.

Kogolla, P., Kisaka, L. G., & Waititu, M. (2004). *ASEI Movement & PDSI Approach.* Paper presented at the Third Country Training in Actitivity, Student, Experiment, Improvisation (ASEI) and Plan, Do, See, Improve (PDSI) Approach, Nakuru, Kenya.

Köller, O., Baumert, J., & Schnabel, K. (2001). Does interest matter? The relationship between academic interest and achievement in mathematics. *Journal for Research in Mathematics Education, 32*(5), 448-470.

Koul, R. B. (2003). *Teacher-student interactions and science classroom learning environments in India.* Unpublished Doctor of Science Education, Curtin University of Technology, Perth.

Krampen, M. (1991). *Children's drawings: Iconic coding of the environment.* New York: Plenum Press.

Kratsios, M. K., & Fisher, D. L. (2003). Cross-cultural family environments of high academic achievers: Parental involvement with early adolescents in the USA, Japan and Greece. In D. L. Fisher & T. Marsh (Eds.), *Making Science, Mathematics and Technology Education accessible to All, Proceedings of the Third Conference on Science, Mathematics and Technology Education, South Africa* (Vol. 1, pp. 303-316). Perth: Curtin University of Technology.

Laborde, C. (1990). Language and mathematics. In P. Nesher & J. Kilpatrick (Eds.), *Mathematics and cognition: A research synthesis by the International Group for the Psychology of Mathematics Education.* Cambridge: Cambridge University Press.

Leder, G. C. (1982). Mathematics achievement and fear of success. *Journal for Research in Mathematics Education, 13*, 124-135.

Leder, G. C. (1987). Teacher student interaction: A case study. *Educational Studies in Mathematics, 122*, 255-271.

Leder, G. C. (1989). Mathematics learning and socialization process. In L. Burton (Ed.), *Girls into maths can go* (pp. 77-89). London: Holt, Rinehart and Winston.

Leder, G. C. (1990a). Gender and classroom practice. In L. Burton (Ed.), *Gender and mathematics: An international perspective* (pp. 9-19). London: Cassell.

Leder, G. C. (1990b). Gender differences in mathematics: An overview. In E. Fennema & G. C. Leder (Eds.), *Mathematics and gender* (pp. 10-26). New York: Teachers College Press.

Leder, G. C. (1990c). Teacher/student interactions in the mathematics classroom: A different perspective. In E. Fennema & G. C. Leder (Eds.), *Mathematics and gender* (pp. 149-168). New York: Teachers College Press.

Leder, G. C. (1992). Mathematics and gender: Changing perspectives. In D. A. Grouws (Ed.), *Handbook of research on mathematics teaching and learning* (pp. 597-622). New York: Macmillan.

Leder, G. C. (1995). Equity inside the mathematics classroom: Class in the world outside the class. In W. G. Secada, E. Fennema & L. B. Adajian (Eds.), *New directions for equity in mathematics education* (pp. 209-224). New York: Cambridge University Press.

Leder, G. C. (1996). Equity in the mathematics classroom; beyond the rhetoric. In L. H. Parker, L. J. Rennie & B. J. Fraser (Eds.), *Gender, science and mathematics: shortening the shadow*. Dordretch, The Netherlands: Kluwer.

Leder, G. C., & Fennema, E. (1990). Gender differences in mathematics: A synthesis. In E. Fennema & G. C. Leder (Eds.), *Mathematics and gender* (pp. 188-199). New York: Teachers College Press.

Leder, G. C., Forgasz, H. J., & Solar, C. (1996). Research and intervention programs in mathematics education: A gendered issue. In A. Bishop, K. Clements, C. Keitel, J. Kilpatrick & C. Laborde (Eds.), *International handbook of mathematics education, Part 2* (pp. 945-985). Dordrecht: Kluwer Academic Publishers.

Leder, G. C., Pehkonen, E., & Torner, G. (2002). Setting the scene. In G. Leder, C, E. Pehkonen & G. Torner (Eds.), *Beliefs: A hidden variable in mathematics education?* (pp. 1-10). Dordrecht: Kluwer Academic Publishers.

Leedy, M. G., LaLonde, D., & Runk, K. (2003). Gender equity in mathematics: Beliefs of students, parents, and teachers. *School Science and Mathematics, 103*(6), 285-292.

Lioong Pheow Leung Yung, M. D. (1998). *An investigation of the teaching methods in mathematics at secondary level.* Mauritius Institute of Education/ University of Mauritius.

Lockheed, M. E. (1985). Some determinants and consequences of sex segregation in the classroom. In L. C. Wilkinson & C. B. Marrett (Eds.), *Gender differences in classroom interaction* (pp. 167-184). New York: Academic Press.

Ma, X. (1999). Dropping out of advanced mathematics: The effects of parental involvement. *Teachers College Record, 101*(1), 60-81.

Ma, X., & Kishore, N. (1997). Assessing the relationship between attitude towards mathematics and achievement in mathematics: A meta-analysis. *Journal of Research in Mathematics Education, 28*(1), 26-47.

Ma, X., & Xu, J. (2004). Determining the causal ordering between attitude toward mathematics and achievement in mathematics. *American Journal of Education, 110*(3), 256-280.

Mahony, P. (1985). *Schools for boys? Co-education reassessed.* London: Hutchinson.

Malone, J. A., & Taylor, P. C. S. (Eds.). (1993). *Constructivist interpretations of teaching and learning mathematics.* Perth: National Key Centre for School Science and Mathematics, Curtin University of Technology.

Mathison, S. (1988). Why triangulate? *Educational Researcher, 17*(2), 13-17.

Matters, G., Allen, R., Gray, K., & Pitman, J. (1999). Can we tell the difference and does it matter? Differences in achievement between girls and boys in Australian senior secondary education. *The Curriculum Journal, 10*(2), 283-302.

Mauritius education and training sector. (2003). Africa Human Development.

McCormick, T. M. (1994). *Creating the nonsexist classroom: A multicultural approach.* New York: Teachers Colloege Press.

McRobbie, C. J., & Fraser, B. J. (1993). Associations between student outcomes and psychosocial science environment. *Journal of Educational Research, 87,* 78-85.

Mead, M., & Metraux, R. (1957). Image of the scientist among high school students: A pilot study. *Science, 126,* 384-390.

Merriam, S. B. (1988). *Case study research in education: A qualitative approach.* San Francisco: Jossey-Bass.

Meyer, M. R., & Koehler, M. S. (1990). Internal influences on gender differences in mathematics. In E. Fennema & G. C. Leder (Eds.), *Mathematics and gender* (pp. 60-95). New York: Teachers College Press.

Middleton, J. A., & Spanias, P. A. (1999). Motivation for achievement in mathematics: Findings, generalizations, and criticisms of the research. *Journal for Research in Mathematics Education, 30*(1), 65-88.

Miles, M. B., & Huberman, A. M. (1994). *Qualitative data analysis.* London: Sage Publications.

Monitoring Learning Achievement. (2003). *A survey of 9 year old chidren in the Republic of Mauritius.* Reduit: Research and Development Section, Mauritius Examinations Syndicate.

Mooney, E. S., & Thornton, C. A. (1999). Mathematics attribution differences by ethnicity and socioeconomic status. *Journal of Education for students placed at risk, 4*(3), 321-332.

Moos, R. H. (1979). *Evaluating educational environments: Procedures, measure, findings and policy implications.* San Francisco: Josey-Bass.

Moos, R. H., & Trickett, E. J. (1974). *Classroom environment scale manual.* Palo Alto, California: Consulting Psychologists Press.

Morgan, C. (1999). Communicating mathematically. In S. Johnston-Wilder, P. Johnston-Wilder, D. Pimm & J. Westwell (Eds.), *Learning to teach mathematics in the secondary school.* London: RoutledgeFalmer.

NCTM. (1989). *Curriculum and evaluation standards for school mathematics.* Reston, VA: National Council of Teachers of Mathematics.

NCTM. (2000). *Principles and standards for school mathematics.* Reston, VA: National Council of Teachers of Mathematics.

Newquist, C. (1997). *The Yin and Yang of Learning: Educators seek solutions in single-sex education.* Retrieved 28 August 2002, from http://www.education-world.com/a_curr/curr024.shtml

Nickson, M. (2000). *Teaching and learning mathematics: A teacher's guide to recent research and its application.* New York: Continuum.

Nunes, T., & Bryant, P. (Eds.). (1997). *Learning and teaching mathematics: An international perspective.* Hove, East Sussex: Psychology Press.

Oakes, J. (1990). Opportunities, achievement and choice: women and minority students in science and mathematics. *Review of research in education, 16,* 153-222.

O'Conner, S. A., & Miranda, K. (2002). The linkages among family structure, self-concept, effort and performance on mathematics achievement of American High School students by race. *American Secondary Education, 31*(1), 72-95.

Olson, V. E. (2002). *Gender differences and the effects of cooperative learning in college level mathematics.* Curtin University of Technology, Perth.

Opdenakker, M. C., & Van Damme, J. (2001). Relationship between school composition and characteristics of school process and their effect on mathematics achievement. *British Educational Research Journal, 27*(4), 407-432.

Parker, L. J., Rennie, L., & Fraser, B. J. (Eds.). (1996). *Gender, science and mathematics: Shortening the shadow.* Dordrecht, The Netherlands: Kluwer.

Pedersen, J. E., & Thomas, J. A. (1999). *Draw-A-Science-Teacher checklist: Children's perceptions of teachers teaching science.* Retrieved 28 April 2002, from http://www2.tltc.edu/thomas/conference%20paper/narst99/narst99.htm

Peterson, P. L. (1988). Teaching for higher-order thinking in mathematics: The challenge for the next decade. In D. A. Grouws, T. J. Cooney & D. Jones (Eds.), *Perspectives on research on effective mathematics teaching* (Vol. 1, pp. 2-26). Reston, Virginia: National Council of Teachers of Mathematics.

Picker, S. H., & Berry, J. S. (2000). Investigating pupils' image of mathematicians. *Educational Studies in Mathematics, 43*(1), 65-94.

Picker, S. H., & Berry, J. S. (2001). Your students' images of mathematicians and mathematics. *Mathematics Teaching in the Middle School, 7*(4), 202-208.

Picker, S. H., & Berry, J. S. (2002). The human face of mathematics: Challenging misconceptions. In D. Worsely (Ed.), *Teaching for depth: Where math meets humanities* (pp. 50-60). New York: Heinemann.

Pintrich, P. R., & Schunk, D. H. (2002). *Motivation in education: Theory, research and applications.* Upper Saddle River, NJ: Merrill Prentice Hall.

Posamentier, A. S., & Stepelman, J. (2002). *Teaching secondary mathematics: Techniques and enrichment units.* New Jersey: Merrill Prentice Hall.

Prithipaul, D. (1976). *A comparative analysis of French and British Colonial policies of education in Mauritius (1735-1889).* Port-Louis, Mauritius: Imprimerie Ideal.

Rajcoomar, V. N. (1985). *A survey of private tuition in primary schools in Mauritius.* Unpublished dissertation for Post Graduate Certificate in Education, Mauritius Institute of Education.

Ramdoyal, R. D. (1977). *The development of education in Mauritius 1710-1976.* Reduit, Mauritius: Mauritius Institute of Education.

Ramma, Y. (2001). *A critical analysis of the performance of girls in physics at upper secondary level in Mauritius.* Unpublished Masters thesis, University of Brighton.

Rao, N., Moely, B. E., & Sachs, J. (2000). Motivational beliefs, study strategies, and mathematics attainment in high - and low - achieving Chinese secondary school students. *Contemporary Education Psychology*(25), 287-316.

Raty, H., Vanska, J., Kasanen, K., & Karkkainen, R. (2002). Parents' explanations of their child's performance in mathematics and reading: A replication and extension of Yee and Eccles. *Sex Roles, 46*(3/4), 121-128.

Rawnsley, D. G. (1997). *Associations between classroom learning environments, teacher interpersonal behaviours and student outcomes in secondary mathematics classrooms.* Unpublished doctoral thesis, Curtin University of Technology, Perth, Western Australia.

Reynolds, C. (2004). Complex learning processes for girls and boys. *Orbit, 34.*

Ria, H., Fraser, B. J., & Rickards, T. (1997). *Interpersonal teacher behaviour in chemistry classes in Brunei Darussalam's secondary schools.* Paper presented at the Conference for Innovations in Science and Mathematics Curricula, Brunei.

Rickards, T., den Brok, P., Bull, E. A., & Fisher, D. (2003). *Predicting students' views of the classroom: A Californian perspective.* Paper presented at the WAEIR Forum, Perth.

Rickards, T., den Brok, P., & Fisher, D. (2003). *What does the Australian teacher look like? Australian typologies for teacher-student interpersonal behaviour.* Paper presented at the Western Australian Institute for Educational Research Forum 2003.

Rickards, T., Ria, H., & Fraser, B. J. (1997). *A comparative study of teacher-student interpersonal behaviour in Brunei and Australia.* Paper presented at the Conference for Innovations in Science and Mathematics Curricula, Brunei.

Rittle-Johnson, B., & Siegler, R. S. (1998). The relationship between conceptual and procedural knowledge in learning mathematics: A review. In C. Donlan (Ed.), *The development of mathematical skills* (pp. 75-110). East Sussex, UK: Psychology Press.

Robertson, H. J. (1998). *No more teachers, no more books: The commercialization of Canada's schools.* Toronto: McClelland & Steward Inc.

Romberg, T. A., & Carpenter, T. P. (1986). Research on teaching and learning mathematics: Two disciplines of inquiry. In M. C. Wittrock (Ed.), *Handbook of research on teaching* (pp. 850-873). New York: Macmillan.

Roscoe, R. D., & Chi, M. T. (n.d). *The influence of the tutee in learning by peer tutoring.* Retrieved 26 May 2005, from http://www.pitt.edu/~chi/papers/TuteeInfluenceFinal_1.pdf

Rowe, K. (1990). What are the benefits of single-sex maths classes? *SET, 1*(9).

Sam, L. C., & Ernest, P. (1999). *Public image of mathematics.* Retrieved 25 May 2005, from http://www.ex.au.uk/~PErnest/pome11/art6.htm

Sanders, J., & Peterson, K. (1999). *Closing the gap for girls in math-related careers.* Retrieved 4 January 2004, from http://www.naesp.org/ContentLoad.do?contentId=474

Savripene, M. A. (2004). La femme est en mesure d'occuper la place qu'elle mérite. *L'Express.*

Saxe, G. B., Gearhart, M., & Nasir, N. S. (2001). Enhancing students' understanding of mathematics: A study of three contrasting approaches to professional support. *Journal of Mathematics Teacher Education, 4,* 55-79.

Schiefele, U., & Csikszentmihalyi, M. (1995). Motivation and ability as factors in mathematics experience and achievement. *Journal for Research in Mathematics Education, 26*(2), 163-181.

Schoenfeld, A. H. (2002). Making mathematics work for all children: Issues of standards, testing and equity. *Educational Researcher, 31*(1), 13-25.

Seegers, G., & Boekaerts, M. (1996). Gender-related differences in self-referenced cognitions in relation to mathematics. *Journal for Research in Mathematics Education, 27*(2), 215-240.

Silver, E. A., Smith, M. S., & Nelson, B. S. (1995). The QUASAR project: Equity concerns meet mathematics education reform in the middle school. In W. G. Secada, E. Fennema & L. B. Adajian (Eds.), *New directions for equity in mathematics education* (pp. 9-56). New York: cambridge University Press.

Sprigler, D. M., & Alsup, J. K. (2003). An analysis of gender and the mathematical reasoning ability sub-skill of analysis-synthesis. *Education, 123*(4), 763-769.

Stevens, T., Olivarez Jr, A., & Hamman, D. (2005). *The role of cognition, motivation, and emotion in explaining the mathematics achievement gap between hispanic and while students.* Retrieved 25 May 2005, from http://www.eduttu.edu/aera/2005/AERA_math.pdf

Stienbring, H., Bussi, B. M. G., & Sierpinska, A. (Eds.). (1998). *Language and communication in the mathematics classroom.* Reston, VA: National Council of Teachers of Mathematics.

Strauss, A., & Corbin, J. (1998). *Basics of Qualitative Research.* London: Sage Publications.

Sumida, M. (2002). *Can post-modern science teachers change modern children's images of sciences?* Paper presented at the Australasian Science Education Research Association Conference, Townsville, Queensland.

Tartre, L. A. (1990). Spatial skills, gender and mathematics. In E. Fennema & G. C. Leder (Eds.), *Mathematics and gender* (pp. 27-59). New York: Teachers' College Press.

Telfer, J. A., & Lupart, J. (2001). *Gender, Grade and achievement differences in student perceptions of parental support.* Paper presented at the Canadian Society for Studies in Education, Quebec, Canada.

Tengar, C. (2001). *An analysis of the gender imbalance at the level of the intake of primary teacher training over a 15-year period in Mauritius.* Unpublished Masters thesis, University of Brighton.

Tiedemann, J. (2000). Gender-related beliefs of teachers in elementary school mathematics. *Educational Studies in Mathematics, 41*, 191-207.

Tobias, S. (1993). *Overcoming math anxiety.* New York: W.W.Norton.

Treagust, D. F., Duit, R., & Fraser, B. J. (Eds.). (1996). *Improving teaching and learning in science and mathematics.* New York: Teachers College Press.

UNESCO. (1990). *World Declaration on Education for All.* Retrieved 23 August 2003, from http://www.cies.ws/PaperDocuments/PDF/WorldDeclarationonEducatio nForAll.pdf

Vale, C., Forgasz, H., & Horne, M. (2004). Gender and mathematics. In B. Perry, G. Anthony & C. Diezmann (Eds.), *Research in mathematics education in Australasia 2000-2003* (pp. 75-100). Australia: Post Pressed Flaxton.

Van den Heuvel-Panhuizen, M. (2005). The role of contexts in assessment problems in mathematics. *For the Learning of Mathematics, 25*(2), 2-15.

Vygotsky, L. (1978). *Mind in society: The development of higher mental processes.* Cambridge , MA: Harvard University Press.

Wakefield, H., & Underwager, R. (1998). The application of images in child abuse investigations. In J. Prosser (Ed.), *Image based research: A sourcebook for qualitative researchers* (pp. 176-194). Bristol, PA: Falmer Press.

Walberg, H. J. (1968). Teacher personality and classroom climate. *Psychology in the school, 5*, 163-169.

Weiner, B. (1971). *Perceiving the causes of success and failure.* New York: General Learning Press.

Willis, S. (1989). *Real girls don't do maths: Gender and the construction of privilege*. Geelong, Victoria: Deakin University Press.

Wolleat, P. L., Pedro, J. D., Becker, A. D., & Fennema, E. (1980). Sex differences in high school students' causal attributions of performance in mathematics. *Journal for Research in Mathematics Education, 11*, 356-366.

Woolfolk, A. (1998). *Educational psychology*. Boston: Allyn and Bacon.

Wubbels, T. (1993). *Teacher-student relationships in science and mathematics classes*. Perth: Curtin University of Technology.

Wubbels, T., & Brekelmans, M. (1998). The teacher factor in the social climate of the classroom. In B. J. Fraser & K. G. Tobin (Eds.), *International handbook of science education* (pp. 565-580). Dordrecht: Kluwer Academic Press.

Wubbels, T., & Levy, J. (Eds.). (1993). *Do you know what you look like: Interpersonal relationships in education*. London: Falmer Press.

Yarrow, A., Millwater, J., & Fraser, B. J. (1997). Improving university and primary school classroom environments through preservice teachers' action research. *International Journal of Practical Experiences in Professional Education, 1*(1), 68-93.

Zevenbergen, R. (2001). Language, social class and underachievement in school mathematics. In P. Gates (Ed.), *Issues in mathematics teaching*. London: RoutledgeFalmer.

Zevenbergen, R., & Ortiz-Franco, L. (Eds.). (2002). *Equity and Mathematics Education* (Vol. 14): Mathematics Education Research Journal.

Wissenschaftlicher Buchverlag bietet

kostenfreie

Publikation

von

wissenschaftlichen Arbeiten

Diplomarbeiten, Magisterarbeiten, Master und Bachelor Theses
sowie Dissertationen, Habilitationen und wissenschaftliche Monographien

Sie verfügen über eine wissenschaftliche Abschlußarbeit zu aktuellen oder zeitlosen
Fragestellungen, die hohen inhaltlichen und formalen Ansprüchen genügt,
und haben **Interesse an einer honorarvergüteten Publikation?**

Dann senden Sie bitte erste Informationen über Ihre Arbeit per Email
an info@vdm-verlag.de. Unser Außenlektorat meldet sich umgehend bei Ihnen.

VDM Verlag Dr. Müller Aktiengesellschaft & Co. KG
Dudweiler Landstraße 125a
D - 66123 Saarbrücken

www.vdm-verlag.de